Full-Contact Leadership

FULL-CONTACT
LEADERSHIP

Edward Flood

Anthony Avillo

> **Disclaimer**
>
> The recommendations, advice, descriptions, and methods in this book are presented solely for educational purposes. The authors and the publisher assume no liability whatsoever for any loss or damage that results from the use of any of the material in this book. Use of the material in this book is solely at the risk of the user.

Copyright © 2017 by
PennWell Corporation
1421 South Sheridan Road
Tulsa, Oklahoma 74112-6600 USA
800.752.9764
+1.918.831.9421

sales@pennwell.com
www.FireEngineeringBooks.com
www.pennwellbooks.com
www.pennwell.com

Director: Matthew Dresher
Managing Editor: Mark Haugh
Production Manager: Sheila Brock
Production Editor: Tony Quinn
Cover Designer: Charles Thomas

Cover image: North Hudson fire chief walks through a tangle of hoselines and debris in the aftermath of a four-alarm fire that destroyed a row of commercial and residential buildings. The fire was one of three simultaneous multiple-alarm fires fought through the night and early morning hours in North Hudson, New Jersey. Author photo courtesy of Ron Jeffers.

Library of Congress Cataloging-in-Publication Data

Names: Flood, Edward, author. | Avillo, Anthony, author.
Title: Full-contact leadership / Edward Flood, Anthony Avillo.
Description: Tulsa, OK : PennWell, [2017] | Includes index.
Identifiers: LCCN 2017007777 | ISBN 9781593703981
Subjects: LCSH: Leadership. | Fire departments--United States--Management. | Communication in management.
Classification: LCC HD57.7 .A945 2017 | DDC 658.4/092--dc23

All rights reserved. No part of this book may be reproduced, stored in a retrieval system, or transcribed in any form or by any means, electronic or mechanical, including photocopying and recording, without the prior written permission of the publisher.

Printed in the United States of America

2 3 4 5 6   21 20 19 18 17

Contents

Foreword by Thomas Von Essen..................xiii
    Imprimatur xv
Preface................................... xvii
Mission Statement........................... xxiii
Eating an Elephant xxv
Acknowledgments xxix
    Chief Flood's acknowledgments xxix
        In memoriam—line-of-duty deaths........... xxxiv
    Deputy Chief Avillo's acknowledgments xxxv
Introduction............................... xli
    Terminology xliv
    As Avillo sees it........................... xlvii
    Basic functions of a leader xlviii

CHAPTER 1

Congratulations on Your Promotion! Now What?............ 1
    The Formal and Informal Organizations 3
    Leadership in the Fire Service 6
        Be consistent................................ 6
        Be honest 6
        Be emotionally mature 7
        Be fair..................................... 7
        Be competent 8
        Be organized................................ 8
        Be intellectually curious 13
        Be respectful of the power of orders 13
    Conclusion 13

CHAPTER 2

Full-Contact Leadership 15
    Introduction to Full-Contact Leadership............... 15
    The Full-Contact Leader as Transformational Tool 19
    The Full-Contact Leader as Servant-Leader 20
    As Avillo Sees It: Weak-Kneed Leadership 21

CHAPTER 3

Basic Styles and Functions of Leadership.................... 25
Autocratic Leadership................................ 25
Participative Leadership 28
Laissez-Faire Leadership 29
Democratic Leadership 30

CHAPTER 4

Group Dynamics and Maslow's Hierarchy of Needs 33
Group Dynamics.................................... 34
Leadership Roles.................................... 34
Maslow's Hierarchy of Needs......................... 35

CHAPTER 5

Rank Has Its Privileges 37

CHAPTER 6

The Dichotomy of Comfort 45
Prologue 45
As Avillo Sees It.................................... 47

CHAPTER 7

Power 57
Personalized versus Socialized Power Orientation 58
  Personalized power orientation 58
  Socialized power orientation 59
Formal and Informal Power 62
Seven Types of Power 63
Leadership Styles and Types of Power 66
The Peter Principle and the Fire Service 68
Proactive and Reactive Application of Power............ 71

CHAPTER 8

Communication.. 73
Verbal Communication.............................. 76
Nonverbal Communication........................... 77
Written Communication 78

Visual Communication 78
Hard-Environment Communication 80
Words at Work 81
Fireground versus Soft-Environment Communication 84
Transactional Analysis............................... 85
    Parent ego state 86
    Child ego state 86
    Adult ego state 86
    Using transactional analysis 86
Case Study: Transactional Analysis the Hard Way—
    As Avillo Saw It................................ 87

CHAPTER 9

Discipline 91
Discipline Is Teaching and Learning 91
Discipline Gets a Bad Rap 92
Three Types of Discipline 93
    Self-discipline 93
    Positive discipline............................... 93
    Negative discipline 95
Discipline Is a Product of Communication 97
You Know It When You See It 99

CHAPTER 10

Blocks to Effective Leadership 101
Leadership Thinking 101
Cognition and Cognitive Premeditation 102
Unintended Consequences........................... 103
Failure of Imagination 104
Functional Fixity................................... 104
Deviation Amplification 104
You Know It When You See It: Failure of Leadership,
    Failure of Imagination, Functional Fixity, Deviation
    Amplification, and Unintended Consequences 105
    Scenario 105
    After–action analysis 107

CHAPTER 11
Casual and Sensual Leadership 109

CHAPTER 12
Absentee and Ambush Leadership 117
    Absentee Leadership 117
    You Know It When You See It 120
    Ambush Leadership................................... 124
    You Know It When You See It: Something Real That
        Might Have Happened 127

CHAPTER 13
Morale and Motivation 131
    Prologue 132
    The Hawthorne Experiments 136
        Hawthorne experiment findings.................. 137
        Hawthorne experiment conclusions 138
    Minneapolis Gas Company Employee Motivation Study... 140
    Morale, Motivation, and the Real World 140
    As Avillo Sees It: The Buy-In 143

CHAPTER 14
Delegation... 147
    As Avillo Sees It.................................... 147
    Introduction to Delegation 147
    More as Avillo Sees It 149
    As Avillo Sees It: The Gift of Imperfection 151
    Guidelines for Effective Delegation.................... 152
    Final Word 153

CHAPTER 15
Setting Expectations 155
    As Avillo Sees It: A Mini Case Study 158
    Establishing Command Presence and Taking Command
        Action 158
    As Avillo Sees it..................................... 160
    More As Avillo Sees It............................... 161
    Avillo Gets Promoted 163

CHAPTER 16

Coaching and Counseling 167
Leader-Coach 172
Leader-Counselor 174
Case Study: Something Real That Might Have Happened . . 180
    The beginning of a comeback 181
    The comeback: righting the wrong. 182
    Expanding the sphere of influence 184
    Some other thoughts on the subject. 185
    Transfer defined 187
    Discipline vs. transfer as punishment 188
    Fun facts regarding transfer as punishment 189

CHAPTER 17

Case Study: The Senior Guy 191

CHAPTER 18

Case Study: The Powers Preinspection—Part I 203
You Know It When You See It 203
Meeting with Division Chief Strong 203
    Leadership guidance. 204
    Overview 205
Inspection Assignment. 206
After-Action Analysis 208

CHAPTER 19

Case Study: The Powers Preinspection—Part II 211
You Know It When You See It: Preinspection Preparation. . 211
Battalion Chief Powers's Skills List. 213
After-Action Analysis 214
Final Word 214

CHAPTER 20

Case Study: The Powers Preinspection—Part III. 217
Inspection of Ladder 20 217
Inspection of Engine 16/Ladder 22. 217
After-Action Analysis 218

Battalion Chief Powers's Post-Inspection Investigation..... 219
    Remedial actions 220
    Instructor training 221
Battalion Chief Powers's Skills List. 225

CHAPTER 21

Case Study: First-Day Experience 227
    You Know It When You See It 227
    Part I: Probationary Firefighter Jane Asher's Orientation . . . 227
        Scenario 227
        After-action analysis. 231
        Overview 231
        Captain Rite's checklist 231
    Part II: Probationary Firefighter Johnny Newton's
        Orientation 233
        Scenario 233
        After-action analysis 234

CHAPTER 22

Case Study: Ladder 13/32 237
    You Know It When You See It 237
    Apparatus Changeover: Ladder 13 to Ladder 32 237
        Shift change 238
        Remedial actions 239
        Overview 241
    Final Word 243
    Unintended Consequences. 243
Postscript. 243

CHAPTER 23

Case Study: The Myth of the Great Firefighter 157
    You Know It When You See It 245
    Something Real That Might Have Happened 246
        Scenario 246
        Commentary. 247
    No Leader, No Follower 248
        Scenario (continued). 248
        Commentary. 249

> Commentary.................................250
> After-Action Analysis 252

Our Last Word(s): Chief Flood's Perspective Parable............253

Index...255

Foreword

Arriving from probationary fire school in July 1970 at Ladder 42, on Prospect Avenue in the South Bronx, was a surreal experience. Fire duty was as good as it could get during the "war years" for the Fire Department of New York (FDNY), which are generally considered to be 1965–75, when FDNY was at its busiest. The firefighters were phenomenal, and the officers were outstanding.

Maybe the officers, through my eyes, were too good. They set a standard for me that I would always try to find in other units. Very often I failed.

My entire career at FDNY spanned almost 32 years. I led the Uniformed Firefighters Association as president and the department as commissioner. When I became commissioner for Mayor Giuliani in 1996, I seized the opportunity and worked to improve leadership for all our firefighters, not just those fortunate enough to be assigned to one of our better units (fig. F–1). I believed then—and am even more convinced now—that accountability and measuring performance are two of the most critical components to develop exceptional leaders.

If you believe it is the responsibility of leaders to be proactive and truly lead, then you will enjoy this book. Chiefs Flood and Avillo are fire service leaders who do not sit by and wait for something to happen. They emphasize that with responsibility comes accountability. The weak-kneed officer is described in one of the many realistic scenarios they present. Taking an oath and accepting the extra pay, in their view, means not closing your office door. The office should not be a hideout! They lay out leadership concepts that are practical. Real-life case stories help the reader to understand what leadership is all about. This includes sharing a clear vision and providing information, knowledge, and methods to realize that vision. The scenarios show you, the reader, ways to lead and balance the often-conflicting interest of those involved. While this may sound like Leadership/Management 101, the scenarios that Chiefs Flood and Avillo describe are from *real life*, not merely academic exercises. They give practical clues on how to deal with problems that face fire service leaders on every tour.

xiv Full-Contact Leadership

Fig. F–1. Leadership forged in fire. *From left:* Lieutenant Kevin "Seamus" Malley, FDNY Commissioner Thomas Von Essen, FDNY Chief of Department Peter Ganci. (Courtesy Kevin Malley)

In a perfect fire department, every leadership position would be filled with a real leader. However, we don't live or work in a perfect world. Therefore, to improve on an imperfect work environment, we need to develop leaders who believe in personal responsibility and accountability. We need leaders who lead, who make decisions that may not always be perfect but are based on their best intentions and knowledge at the time.

> *Reading and studying the lessons provided by Chiefs Flood and Avillo will help you to improve at every leadership level. They get it! You can too! When you take that oath, take it seriously, be proud, and use this book on your leadership journey.*
>
> —Thomas Von Essen

Imprimatur

Tom Von Essen was an FDNY firefighter, the president of the firefighters' union and he was Fire Commissioner of FDNY on September 11, 2001. Any one of the positions listed would have provided a couple of lifetimes' worth of full-contact leadership experience.

An *imprimatur* is an official thumbs-up, a seal-of-approval issued by some authority. Originally *imprimatur* literally meant a license to print. The authors are awed that Commissioner Von Essen would give his substantial imprimatur declaring *Full-Contact Leadership* fit to print. Thank you.

Preface

Fire service leadership is not a position; it is a stewardship.
—Flood

Knowledge is something. Knowledge properly applied is something else.
—Flood

Fire service leaders must keep in mind that they are only holding a place that belongs to a next generation of leaders. The fire service leader should actively work to enable and create the leaders who will take their place.
—Flood

Know it. Understand it. Effectively do it. Evaluate it. Create something new and better out of it.
—Flood

Leadership and learning are indispensable to each other.
—John F. Kennedy

Don't find fault, find a remedy.
—Henry Ford

The quality of a leader is reflected in the standards they set for themselves.
—Ray Kroc

I suppose leadership at one time meant muscles; but today it means getting along with people.
—Mahatma Gandhi

The task of the leader is to get his people from where they are to where they have not been.
—Henry A. Kissinger

Leadership is the capacity and the will to rally men and women to a common purpose and the character which inspires confidence.
—General Bernard Montgomery

Knowledge is the ability to recall facts, terms, basic concepts, and other information regarding a specific subject. Knowledge (rote memory) is the lowest level of learning.

Knowing something or knowing something about something does not guarantee that knowledge can be applied. It doesn't even guarantee that the information is understood (comprehended).

Comprehension is the ability to grasp information intellectually. Comprehension is an understanding of a central idea. Comprehension is the level of learning that allows information to be absorbed, interpreted, and translated into action.

Comprehension is the next-lowest level of learning. Understanding information is a step up from just knowing about a subject. However, comprehension does not guarantee an ability to apply information.

Much of our early educational experience placed emphasis on knowing and comprehending. In the fire service, knowledge and comprehension are of little use if the information cannot be applied and translated into effective action. To be applied, knowledge and comprehension must be married and put to use in real world situations.

Knowledge and comprehension are about the theoretical. Application is about the empirical.

Application is where knowledge and comprehension go when they grow up. Application is the ability to use acquired knowledge. Application is about solving problems and applying acquired knowledge, facts, principles, skills, techniques, and rules in new, effective, and creative ways.

Analysis is the next-higher level of learning. Analysis allows for the examination and breakdown of information and actions into parts. Analysis identifies the modus, the cause and the effect of actions. Analysis identifies the rationale and logic that can support theories, proofs, and generalizations.

Synthesis is the highest form of learning. Synthesis brings together knowledge, information, theories, principles, skills, and applications to create a new paradigm, template, solution, process, or methodology. Synthesis combines diverse elements to create an alternative solution.

Full-Contact Leadership encourages all those who envision themselves to be leaders in the fire service to aspire to embrace every level of learning.

The United States fire service is the preeminent firefighting force in the world. The fire service is society's Swiss Army Knife.

Whether the job is a room-and-contents fire, an all-hands multiple alarm, a month-long wildfire campaign operation, a chemical tank farm, or a trashcan fire on the corner, every fire that ever started has been confined, controlled, and extinguished by men and women who selflessly extend themselves to perform extraordinary acts under the most extraordinary conditions. Every life that could have possibly been saved has been saved. Every variety of emergency has been mitigated. The preceding is a very short list of what the fire service contributes to society. Even the short list is quite an amazing record of achievement.

The physical laws of combustion are constant and universal. The combustible load, array of contents, structural configuration, environmental conditions, and human intention or folly dictate the path, progression, and form that the natural laws of physics will present and progress. Add to the equation an infinite number of seen and unforeseen variables, and you've got yourself a whole bunch of good reasons for society to have its fire, rescue, and emergency services no further than a phone call away.

Fig. P–1. Leadership is earned. The fire service leader is involved in all facets of the operation, leading from the front, the rear, and everywhere else.
(Courtesy Ron Jeffers)

The linchpin that holds together and harnesses fire, rescue, and emergency operators is the fire service leader.

Leadership in the fire service is an awesome responsibility and one of the most rewarding experiences that life has to offer.

Full-Contact Leadership is written for men and women who hold leadership positions or aspire to leadership roles in the fire service.

There are a lot of leadership positions in the fire service. Not every leadership position is filled by a leader. Leadership has very little to do with the color of your helmet, the bling on your collar, the stripes on your sleeve, the title on your door, the order of march, or the crease in your pants. Leadership is a commitment to drawing out the very best within others and allowing the very best in others to be expressed as excellence. Leadership is a career-long, ever-challenging, never-ending, self-initiated, self-sustained personal research, development, and improvement program. Leadership is never about "you"; it's always about "them."

Leadership is the greatest asset of any fire department. There is nothing good happening in the fire-rescue world without effective leadership.

There is no such thing as wishing for leadership.

Every fire department has all the raw material necessary to produce an assembly line of effective leaders.

In order to properly serve the community, the profession and the men and women who operate at fires, rescues, and emergencies must create, support, and maintain a culture that values effective, creative, committed, and employee-centered leadership above everything else.

Within the fire service there is no more important or valuable asset than an effective leader. Fire departments train the hell out of their firefighters. At the same time there exists a chronic underinvestment of time, money, effort, and energy when it comes to leadership development and training. The effort extended versus the reward potential is often way too inadequate.

Firefighting demands great physical, mental, and emotional effort. Give the fire service a job that can be handled with manpower, womanpower, tools, equipment, blood, sweat, tears, and all the rest, and everyone knows that the job will be successfully completed post-haste. Fire operators are great with things they can hold, throw, open, close, force, respond to, extinguish, control, save, and on and on. Leading men and women in the performance of these tasks presents challenges that go deep into the world of the human condition. A fire service leader must be learned enough to contend with the infinite variety of human behavior, folly, fear, and courage.

As the fire service recognizes, prepares, and meets the complexities found on the fire and emergency ground, the fire service must recognize and prepare for the social, psychological, emotional, and humanities-based landscapes of the 21st century.

Leadership is not an accident of birth. Babies are born. Leaders are built.

I have always been skeptical of the "born leader" school of thought. For instance, George Armstrong Custer is thought to have been a "born leader."

When all is said and done, when all the theories, philosophies, and principles are discussed, written about, and studied, leadership comes down to choice. Leadership is a choice, pure and simple.

Do I choose to lead or not? This is a question that is implicit in every leadership transaction. Becoming an effective leader demands a deliberate, committed decision to immerse in a career-long program of personal and professional development and improvement. When a conscious, thoughtful decision is made, the aspiring leader must become a rabid and voracious consumer of any and all leadership information available. A leader will emerge when leadership theory becomes leadership applied.

The safest, most effective path of least resistance into the future is through a creative and effective fire service leadership corps. If today's leaders aren't making the grade, the fault and blame must be laid at the feet of the leaders who have come before.

The idea that going to a lot of fires is the crucible for good leadership is just plain wrong. Even the busiest fire departments spend about 5% of their time, energies, and effort on the emergency ground. The other 95% of the time is dedicated to standing down and preparation.

Anybody with 20 years (or any amount of time) on the "JOB" most certainly has experience. The quality of experience and the lessons learned will always be more important than the quantity of experience. It was told to me that "You can have 20 years dealing with and experiencing the same things, the same old way over and over again, or you can have 20 years of learning and growing with each new experience."

Full-contact leadership is leadership that gets "in there" and "gets its hands dirty." Full-contact leadership is prepared, present, engaged, courageous, employee-centered, and mission-focused. Full-contact leadership is not for the faint-hearted.

> **Note:**
>
> This scenarios presented in Full-Contact Leadership are combinations of experience and composite fictions put together to make a point. Unless otherwise cited, names, characters, places, events, and incidents are either the product of the authors' imagination or used in a fictitious manner. Any resemblance to actual persons, living or dead, or actual events is purely coincidental.

Full-Contact Leadership Mission Statement

Full-contact leadership is smart, creative, and interventionist stewardship resourced and equipped to bring to bear management, leadership, supervisory, and communication arts and sciences to effectively mitigate and empirically improve issues, situations, or incidents at the point closest to the challenge.

Full-contact leadership flows from individual and organization-wide commitments to the provision of high-level, quality leadership due the men and women who share perilous duty in the service of others.

Full-contact leadership is 24/7, consistent, non-shirking. Full-contact leaders come to the game with skill portfolios deep and rich enough to enter the uncomfortable zone in order to proactively confront the uncomfortable.

Full-contact leadership is an employee improvement, protection, maintenance, and repair program. Full-contact leadership believes that well-trained, maintained, and protected individual operators are the most effective and productive incident mitigation agents.

Full-contact leadership in the fire service happens in real time, up and down the chain of command at every point where the organization and its mission meets the fire operator.

Full-contact leadership seeks to align the personal needs and personal vision of fire operators with the needs, vision, and mission of the fire service organization.

Full-contact leadership is an in-line, up-and-down-the-line human condition monitoring system allowing preemptive leader interventions.

Full-contact leadership is most effective when organizational commitment is supported and appropriately resourced.

Full-contact leaders are serious people doing serious business while not taking themselves too seriously.

Full-contact leadership *is the highest compliment an organization or a boss can pay to the people they lead.*

Full-contact leadership *more than anything is a continuum of creatively combined managerial arts and sciences introduced with surgical precision onto and into specific situations at the very point of contact between the member and the organizational requirement.*

Eating an Elephant: Operations Chief Wanted

Candidate must be mission-focused, unyielding yet flexible. The applicant must possess the hide of a rhinoceros and also be a "people person." The position will require the candidate to be point of the spear, lightning rod, and agent of change.

The applicant must have the capacity to reconcile, mollify, placate, conciliate, compromise, administer, enforce, educate, and be prepared to eat crow as required. The capacity to tap dance and/or do a mean soft shoe will serve the candidate well.

The candidate will be required to find consensus and creative solutions amid a jungle of power bases, egos, recalcitrance, obstructionism, and a landfill of opinions (some valuable, some not so much).

The applicant must be able to recognize, harness, and employ all the creativity, positive energy, and great potential inherent in every member and the organization as a whole.

The candidate will work under and report to two chiefs of department, two politically appointed, hands-on fire directors, and a board of fire commissioners comprised of the mayors and city managers of the five regionalizing municipalities.

The specification for the position is to combine and organize one volunteer and four career fire departments into a single, cohesive, state-of-the-art, highly effective, twenty-first-century, urban fire, rescue, and emergency service.

The combined department will provide fire, rescue, and emergency services to one of the most diverse and densely populated urban environments in the United States.

If not experienced the candidate must agree to participate in an accelerated, self-administered, trial by fire, "on the job" training program.

This job will provide unique opportunities working with more than 300 firefighters, captains, battalion and deputy chiefs, seven separate unions, seven enforceable union contracts, seven pay scales, two different starting times, five different vacation schedules, a less-than-enthusiastic rank-and-file, as well as a small army of lawyers.

The candidate will need to recognize that each individual fire department brings a separate and unique informal and formal culture to the nascent department. Sensitivity and a capacity to deal effectively with this reality are essential to the successful merger of the entities.

Applicant will be required to unify and standardize alarm response and fireground operations, including command-and-control and communication procedures. The applicant will further be tasked to advise, coordinate, and facilitate all attendant administrative, support, and maintenance demands, issues, and activities.

Prior to consideration, the applicant shall create and submit for review and approval:

1. A plan and timeline detailing step-by-step the methods and actions proposed to bring about the successful accomplishment of the expressed mandates.
2. A comprehensive and functional manual of rules and regulations encompassing all roles, responsibilities, and codes of conduct as applied to every position, rank, and member of the service, including all inter- and intradepartmental assignments.
3. A comprehensive set of standard operating procedures encompassing every action, activity, and aspect of operations necessary for the safe and effective response, mitigation, command, and control at fire, rescue, and emergency operations.
4. The applicant will be required to attend an endless number of "high level" meetings, many of which the applicant's presence will be resented and deemed an encumbrance.

The truth of the matter is that no such want ad ever existed. A second truth is that no one knew or could have known exactly what the specifications for the chief of operations (North Hudson Regional Fire & Rescue) would look like. Regionalization of five separate fire departments was never attempted, let alone made a reality, in the state of New Jersey.

Eating an Elephant: Operations Chief Wanted xxvii

Fig. E–1. Chief Flood consults with an FDNY chief the morning after the North Hudson "Night of Flames." Four 4-alarm fires struck in Union City within a 45-minute time frame, bringing in extensive mutual aid including an FDNY Task Force. (Courtesy Ron Jeffers)

I accepted the chief of operations NHRF&R assignment, and in eight months with the help and support of a long list of contributors, North Hudson Regional Fire & Rescue became a high-functioning reality. Today North Hudson Regional Fire & Rescue is a motivated, aggressive, premier, 21st-century fire, rescue, and emergency service provider.

NHRF&R provides highly effective, professional fire, rescue, and emergency services to a combined population of more than 210,000 residents packed into 10 square miles of the most densely populated urban area in the United States. On a daily basis, tens of thousands of commuters and untold tons of material and goods move along and through North Hudson's roadways, waterways, and tunnels in cars, trucks, buses, trains, light rail, subways, helicopters, water taxis, and ferries.

Is there anything in the world better than being a firefighter?

I'm goin' with nope.

Edward Flood
Chief of Department (Ret.)
North Hudson Regional Fire & Rescue

Fig. E–2. North Hudson, New Jersey. There is a never a fire without several exposure problems and it is always rush hour. When the buildings outnumber the trees, things can get pretty exciting.

Acknowledgments

Chief Flood's Acknowledgments

The "JOB" makes the man.

—Flood (fig. A–1)

Fig. A–1. Chief Ed Flood on the job in 2000 (Courtesy Ron Jeffers)

This was the hardest section of the book to write. This is a thank-you note to every firefighter, fire officer, and chief officer who I have been fortunate enough to cross paths with. There is no way that I could possibly list everyone who played a role in my fire service experience.

For every one name that shows up here, there are scores of others who deserve recognition. A firefighter's career is one long, continuous learning experience. For good, bad, ugly, or indifferent, everyone in this business has a lesson to teach. So, please accept this report as a heartfelt "all-hands" expression of gratitude, appreciation, and respect.

Let me start with two chief officers who opened my mind to master's- and doctorate-level fire-think: Gerald Huelbig, Deputy Chief, Chief of Operations, Weehawken, New Jersey, Fire Department (retired); and the late Frank Constantinople, Chief of Department, Jersey City Fire Department. Chief Huelbig had wisdom, commitment to mission, and an amazing capacity to subjugate his ego for the good of the Weehawken Fire Department (WFD) and the welfare of the personnel under his command. Chief Huelbig, along with Public Safety Director Jeff Welz, was tasked with breathing life back into a department that was suffering from years of neglect by political office holders, low morale, and organizational malaise. When I was a brand-new ladder-company officer, Chief Huelbig gave me support (often indicated with a thumbs-up, or a thumbs-down when I needed to think twice) and as much leash as I needed to spearhead and shepherd the changes he saw as necessary to revive and modernize an ailing fire department.

Chief Frank Constantinople was a big guy, an ex-marine, a tough and experienced firefighter, officer, and no-nonsense leader. Chief Constantinople was the epitome of hard-nosed. I met Frank when he was a battalion chief teaching a promotional preparation class. Frank possessed encyclopedia-like knowledge of all things fire and carried himself in a manner that loudly announced a command presence.

Taking a page from the good-cop/bad-cop school of mentoring, Frank and Gerald took me under their collective wing. They roughed me up, gave me a little polish, and helped me to develop my own voice and template for the type of leadership I aspired to. I owe them both a great deal, and I thank them.

Learning how to lead is more than anything a bottom-up experience. I cut my leadership teeth as a company officer and learned about being a committed, contributing, and mission-focused leader in the fire service. The firefighters I worked with were the ones who taught me the "secret handshake of leadership," and I was fortunate to work with some of the very best firefighters any officer could wish for.

Rich Barreres was the first probationary firefighter assigned to my command (fig. A–2). He was young and smart, with an endless supply of opinions and attitudes and a real-deal desire to be good at the "JOB." Working with a probationary firefighter is a great responsibility, challenge, and privilege. The probationary firefighter is as much a teacher as he is a student. Richie turned out to be a great firefighter and helped to make me a better officer, and he eventually ended up a rescue captain with North Hudson Regional Fire & Rescue (NHRF&R). Captain Barreres was performing a primary search on the second floor of a residential structure when the fire rolled over into the room and forced Richie to bail out the window. Although Captain Barreres survived, he had to go off the "JOB" due to injuries sustained. For him, having to retire early was more painful than any of his injuries.

Fig. A–2. Battalion Chief Mike Hern and Rescue Captain Richie Barreres. (Courtesy Bill Menzel Photography)

My grandfather, Edward Francis Flood, was an engine company captain with Engine 97, FDNY, and my uncle, Peter Prial, worked in the Bronx as a lieutenant for FDNY. I followed in their footsteps by becoming a firefighter, and today my son Terrence and Richie's son are both newly promoted captains and company commanders. Having a relative or close relation in the service is not a prerequisite to becoming a good or great firefighter. Still, it does seem that once you get a firefighter in your family, it is quite common to find another somewhere on your family tree.

I worked ten years as a ladder officer in a double-company house, Engine 203/Ladder 222. Tim Finnegan was the captain of Engine 203. I was a rocket and he was a rock. Captain Finnegan was an amazing officer—quiet, competent, smart as a whip, and a "first-in/last-out" kind of engine officer. Tim was, without a doubt, the only officer who could've worked in tandem with and put up with my "effusiveness" (to put it politely).

As a company officer, my greatest learning experience occurred when Anthony Avillo and Dave Curtis were assigned to my ladder company. Both Dave and Anthony have risen through the ranks and today serve as deputy chief/division commanders for NHRF&R. Together we made a commitment to tear apart every ladder-company–related IFSTA (the basic training manuals of the International Fire Service Training Association) and to read about, research, and replicate every tool use and ladder operation, as well as improvise every type of equipment construction and device pictured or described in the IFSTA books. The process made us an effective, formidable, and justifiably confident team of ladder company operators. Being a ladder company commander was the most rewarding and fun portion of my 30-year career.

WFD personnel who had an impact on my career include but are not limited to this short list: Deputy Chief Dave Curtis, Deputy Chief Anthony Avillo, Captain Bill Lemonie, Captain Jim McDonough, Captain Richie Barreres, Captain Eddie Conners, Captain Dave Flood, Battalion Chief Mark Lorenz, Captain Joey "Bones" Rodriquez, Captain Bobby Ellebrock, Battalion Chief Steve Quidor, Captain Kevin O'Driscoll, Captain Brian McGorty, Captain Dan Repetti, Captain Dennis Rudd, Captain Jim Stelman, Captain Allen Dembroe, Battalion Chief Frank Nagurka, Battalion Chief George Browne, Public Safety Director Jeff Welz, Captain Andrew Scott, Captain Tim Finnegan, Firefighter Mike Smayda, Deputy Chief "Pop" Pizzuta, Captain George Pizzuta Sr., Battalion Chief Pat Pizzutta, Battalion Chief Victor Vangelokos, Deputy Chief Joe Fredericks, Captain George Pizzuta Jr., Captain Rick Testa, and Firefighter Frank Baker.

Many of the WFD guys listed above went on to company, battalion, and division command positions in NHRF&R, which is the first and only regionalized fire and rescue department in the State of New Jersey. Legions of uniformed and civilian individuals brought vision, expertise, and extraordinary effort together to create and successfully launch NHRF&R. The following is a short and incomplete list of those who helped, supported, and otherwise had an impact on the NHRF&R portion of my career: Battalion Chief Mike "The Doctor" Hern, Director Mike Diorio, Director Jeff Welz, Weehawken Mayor Rich Turner, North Bergen Township Manager Joseph Auriema, Chief Robert Aiello, Linda DiPaolo, Chief Anthony Presutti, Chief Bob Jones, Chief Brion 'Mac'

MacEldowny, Chief Frank Montagne, Deputy Chief Marty Mandell, Deputy Chief Bob Montagne, Deputy Chief Nick Gazzillo, Deputy Chief Mike Falco, Battalion Chief Steve Quidor, Captain Rob Focht, Battalion Chief Angelo Caliente, Battalion Chief Bobby Agostini, Captain Sean Mick, Battalion Chief Ronnie Tompkins, and so many more.

This above oh-so-incomplete list represents only a fraction of the civilians, firefighters, and fire officers who climbed a steep and slippery slope to create one of the most effective and premiere firefighting forces in the State of New Jersey. NHRF&R provides fire, rescue, and emergency services to the most diverse and densely populated urban center in the United States.

I have had the great privilege to be given command of two fire departments: the WFD and NHRF&R. I need to acknowledge and express a special note of gratitude and commendation to every firefighter, captain, and chief officer who performed honorably, aggressively, bravely, and selflessly when I had the privilege of their command.

It is important for me to recognize, congratulate, and commend the amazing efforts extended by every member of the Union City, North Bergen, Weehawken, West New York, and Guttenberg fire departments. The firefighters and officers from these five individual departments, working together to overcome countless issues and difficulties, made regionalization in New Jersey a reality. These firefighters coalesced, formed bonds, and made a commitment to a mission that turned a concept into a reality.

Before I end my gratitude list, I must extend a special note to Battalion Chief Mike "The Doctor" Hern (fig. A–2). In 1994 Mike and I started Study Group inc (SGi), a test preparation and consulting business. SGi was very successful. A great majority of our students went on to become leaders at the company and chief officer levels in fire departments throughout New Jersey. When demand for our services grew beyond the capacity of a two-man operation, Anthony Avillo came on board. Captain Avillo was a talented, young, and hungry company officer and former SGi student—and was a perfect fit. Both Anthony and I will, without hesitation, admit that the Doctor was the force behind any success SGi might have had. Mike Hern served as a model firefighter and company officer with the Union City Fire Department and in 2013 retired as a Battalion Chief, Commander of the 1st Battalion, Division 2, NHRF&R.

In memoriam—line-of-duty deaths

Firefighter Vincent Neglia

Engine 13, Battalion 3, Division 4, NHRF&R

Succumbed to injuries sustained September 9, 2006, when an apartment flashed over while conducting a primary search

Battalion Chief Robert "Bobby" Agostini

Battalion 1, Division 3, NHRF&R

Succumbed to complications of pulmonary injuries sustained while working at Ground Zero, August 20, 2013 (fig. A–3)

Fig. A–3. Battalion Chief Bobby Agostini . . . "We All Shine On" (Courtesy Ron Jeffers)

Deputy Chief Avillo's Acknowledgments

First and foremost, I would like to thank my mom and dad, who taught me lessons in leadership and humility that I still and always will carry with me. Although my dad passed away over 35 years ago, his leadership traits and influence on me resonate to this day, and I am who I am because of his guiding hand. Although not a firefighter, his inspired work and organizational talents in coaching and mentoring kids and parents alike are still well remembered and acknowledged within the community. Not a day goes by that I don't think of him or his love for our family. After we lost him, my mom became the rock that guided three absolutely wild and crazy teenage sons to productive adult lives—not the easiest of tasks. Great job, Mom!

I also need to thank my two great brothers; my precious daughters, Stephanie and Lindsay, who are my main reasons for getting up each day; and my wonderful family for supporting me and sharing me with the "JOB" I love (figs. A–4 and A–5). The best job in the world? You better believe it!

Fig. A–4. Anthony Avillo on the "JOB" in 2014 (Courtesy Ron Jeffers)

Fig. A–5. My dad, Anthony D. Avillo, a builder of youth—and a leader of men

I would also like to acknowledge the leadership and life lessons of the late Coach Vincent Ascolese, who coached North Bergen High School and Hoboken High School football for 50 years. His gift for motivation, wisdom, and guidance influenced countless young men. I am proud to count myself among those who benefited from his mentoring.

Firefighting is a lot like playing football. Both are team oriented and require an unwavering and unquestioned sense of dependency and trust. Your A game must be in place at all times because your fellow firefighters, like your teammates, are depending on you—and you upon them. Both activities require extreme dedication and training. Both drip with adrenaline and excitement. Both are stoked by pride and tradition that only those who participate can understand.

Many of the firefighters and officers in North Hudson had the good fortune to play football under Coach Ascolese. Mr. A. taught many of my fellow firefighters in NHRF&R to set goals, work hard to achieve those goals, and above all else believe in yourself and your team. Coach Ascolese preached that your team was a part of your family—and that family comes first. As NHRF&R Deputy Chief Mike Falco said, "He had a way of reaching into your soul and making you the best you can be." "He taught us to expect to win," added Battalion Chief Mark Johnson. Finally, Captain Nick Prato offered, "Growing up in North Bergen, there were two things you wanted: to be a Bruin and to play for Coach Ascolese."

An opposing coach once said, "Playing North Bergen is like anticipating a hangover. You know what to expect. You just don't know how to stop it." I also remember the phrase Coach Ascolese often used: "In this life, you can do anything you want to do—if you want it bad enough and if you are willing to do the work." His words guided my actions throughout my time as a player and later as a coach at North Bergen High School. The traits that make a champion in football—such as self-discipline, dedication, and reliability—also apply to the way we carry ourselves as firefighters and fire officers. These traits, learned early on the gridiron courtesy of Coach Ascolese and his staff, have stayed with me and helped me be a better person, fire officer, and leader (fig. A–6).

Fig. A–6. North Bergen Coach Vincent Ascolese. If this look was directed at you, you did not even want to come back to the sidelines. (Courtesy of the Ascolese family)

My career has been documented previously in the *Fireground Strategies* textbooks. I spent 30 years as firefighter and officer with WFD and the NHRF&R. I finished my career as a deputy chief and regional tour commander of the "Fighting 1st" Platoon at NHRF&R. I want to acknowledge and thank all those who have worn the uniform of the WFD and the NHRF&R.

I want to thank Professor Kevin "Seamus" Malley for his unending support and belief in me. Professor Malley retired as a lieutenant from Ladder 40, FDNY, in 2001. Seamus was chairperson of New Jersey City University's fire science department. He recently retired as assistant dean of professional studies. His career has been dedicated to creating the best fire officers in the United States. Education is the foundation of the fire service, and the combination of the classroom and solid training and fire ground experience cannot be matched. A firefighter or officer lacking in any of these three attributes cannot yet be considered a complete and well-rounded member of the fire service.

I am currently deputy fire marshal and director of the Monmouth County (NJ) Fire Academy. I would like to acknowledge Fire Marshal Kevin Stout and the Fire Academy staff for their dedication and support. You have recharged my batteries. I also want to acknowledge my colleagues from the Bergen County Fire Academy as well as the countless friends I have met along the way across the country who have made the fire service experience most rewarding.

Further, I would not be in this book business if it weren't for the following people. Thanks to Peter Hodge for starting it all off. I'd like to acknowledge Bobby Halton, Marla Paterson, Diane Rothschild, Mary McGee, Cindy Hughes, Mark Haugh, Eric Schlett, Scott Nelson, Mary Jane Dittmar, Ginger Mendolia, Rob Maloney, Tommie Grigg, Sheila Brock, and Eileen Brennan. Without the support of these people, no one would know my name.

OK. Now about Flood. Ed Flood was not my first captain. That was Tim Finnegan, as knowledgeable, strong, and patient a fire officer as ever was. I was coming off of playing college football and had always wanted to be a firefighter. Basically, I didn't know a screwdriver from a hammer. Actually I wasn't that bad, but I was close. However, I was also young, strong, and willing to learn. As a probie, I approached the job like a sponge, and Captain Finnegan passed on an invaluable education to me that has lasted throughout my career.

I was originally assigned to Engine 203. Across the apparatus floor sat the snarling Ladder 222. On my shift, it was commanded by Captain Ed Flood. Like Ladder 222, he would also snarl at us across from the other side of the apparatus parking lines, with that air of superiority only "truckies" know and can display. I had met him at the gym before I got on the job, so I eventually wanted to be on his rig. My early perception of ladder work was that it was about breaking things, and since I was much better at breaking things than fixing things or putting them back together, I figured ladder work was for me. I was on the engine and had to learn engine work before I was allowed to cross the floor. After a short time on Engine 203, I was transferred over to Ladder 222 (the "Triple Deuce") and began a career-long journey with then Captain Ed Flood (fig. A–7). Captain—and later Chief—Flood was always one rank above me and directly superior, so even if I didn't want to, I was continuously bombarded with the genius and mental meanderings of Ed Flood. There were times when I didn't even have to ask because he would often talk to himself before he would blurt out an idea. He was constantly thinking about and improving the job. He pushed me to reach for the stars, and the most recent star I have reached for is this book.

Fig. A–7. Chief Flood swears in Battalion Chief Avillo as Deputy Chief, 2002. (Courtesy Ron Jeffers)

The most dangerous leadership myth is that leaders are born—that there is a genetic factor to leadership. . . . Leaders are made rather than born.

—Warren G. Bennis

Before you are a leader, success is all about growing yourself. When you become a leader, success is all about growing others.

—Jack Welch

The first responsibility of a leader is to define reality. The last is to say thank you. In between, the leader is a servant.

—Max DePree

Leadership *is the capacity to translate vision into reality.*

—Warren G. Bennis

A leader is best when people barely know he exists. When his work is done, his aim fulfilled, they will say, "We did it ourselves."

—Lao Tzu

You manage things; you lead people.

—Grace Murray Hopper

Management is doing things right; leadership is doing the right things.

—Peter F. Drucker

The greatest leader is not necessarily the one who does the greatest things. He is the one that gets the people to do the greatest things.

—Ronald Reagan

The key to successful leadership is influence, not authority.

—Kenneth H. Blanchard

Good management is the art of making problems so interesting and their solutions so constructive that everyone wants to get to work and deal with them.

—Paul Hawken

"Fini Origine Pendent"

The end depends on the beginning.

Introduction

There are many excellent books on the art, science, and theory of how to supervise and manage human beings. Just as much information is available on leadership and how to lead humans to water and get them to drink. All of these texts speak well to—and explain in depth—the wide field of management theory and its application. Informative resources on management practices, supervision, leadership, and command theory are available to anyone seeking the privilege of being someone else's boss.

We encourage all those who are honored and gifted with leadership positions in the fire service to dig deep into and mine the mountains of information that explore the art, science, and theory of leading, managing, and supervising. Anyone who holds a command-level position in today's fire service must become and remain a career-long student and proficient practitioner of the art and science of management, leadership, and command (fig. I–1).

Fig. I–1. The challenge of command ultimately requires a career-long commitment to learning, leading, and mentoring—under all conditions. Not everyone is cut out for this role. (Courtesy Ron Jeffers)

To carry out supervisory functions, a leader must develop a skill portfolio. This portfolio should contain, at a minimum, communication skills (transactional analysis) and an understanding of the different types of power and leadership styles. Moreover, every leader needs a working knowledge of group dynamics and Maslow's hierarchy of needs.

Achieving rank is meaningless if there is no meaningful contribution. Leadership has been described as a process of social influence in which one person can enlist the aid and support of others in the accomplishment of a common task.

Fortunately, the U.S. fire service is blessed with a legion of fire officers who possess strength of character, solid professional experience, dedication to duty, and a serious sense of responsibility toward the women and men they lead. These fire officers are highly competent, well prepared, and motivated. Daily and nightly, these officers lead the proud men and women of the fire service into hazard-rich environments and then bring them safely home to their families, their loved ones, and their lives (fig. I–2).

Fig. I–2. The ultimate goal of the fire service leader is to bring the troops home.

My leadership experiences unraveled in a nonlinear and asymmetrical fashion. There has rarely, if ever, been a one-size-fits-all answer to any issue or situation I have dealt with as a leader or manager. Successful, effective, and creative leadership demands intellectual curiosity, creativity, constant subjugation of ego, and courage of conviction and conscience—bending but never breaking. To be effective, every fire officer must develop a portfolio full of team-building and management skills. The knowledge and capacity to bring together the elements for successful, creative, and effective leadership do not come cheap.

This book is designed to stand alone as a reference. We have tried to create a book that can easily lend itself to both individual and group study. Essentially, we aimed to create a manual that is fire-leader friendly to make leadership information more digestible for the men and women who do street-level leading in the real-deal world of the soft and hard fire, rescue, and emergency environments (fig. I–3).

Fig. I–3. Leaders of all ranks must be "on" at all times, whether in the soft environment of the nonemergency world or the big stage, real-deal hard-environment world of the fire ground. (Courtesy Ron Jeffers)

Full-Contact Leadership presents information in a variety of ways. There are straightforward chapters explaining concepts, as well as asides, quotations, and case studies. Throughout the book, we have included quotes that apply to the information presented and amplify the text with examples of wisdom. Quotes are small dense packets of information and wisdom. Quotes are bursts of brilliance. A single quote can synthesize an entire chapter of text into one or two sentences. Quote injections serve to immunize the reader from the toxic effects of any intellectual anemia that may be expressed by the "author/experts".

Many chapters are graced with sections titled "As Avillo Sees It" (as well as variations on that theme). With these inclusions, the reader should realize that Chief Flood is grudgingly admitting that there is indeed an audience for Deputy Chief Avillo's fireside insights. These represent an invaluable contribution to this book and are a boon to the entire world of fire science literature.

Case studies, some comprising entire chapters and others, short sections titled "You Know It When You See It," provide the reader a window into the theoretical aspects of leading, managing, and supervising as applied in real-world fire situations. The hypothetical scenarios presented can be considered as "real things that might have happened." Many of the case studies are designed to create dramatic contrast between mildly exaggerated examples of poor leadership and functional models of positive, effective, and full-contact leadership behaviors. At the end of each case study, we invite readers to identify the leadership styles, skills, and tools that are demonstrated in the case studies.

It is recommended that readers of this textbook form study groups. Studying in groups is a powerful and effective learning strategy. We challenge groups of students not only to dissect and critique the information presented but to argue, debate, and brainstorm among themselves.

Terminology

Deputy Chief Anthony Avillo's text *Fireground Strategies* introduced the terms *hard environment* and *soft environment*. Similarly, fire department personnel have occasionally been described as fire service "software," while apparatus, facilities, tools, and equipment are considered fire department "hardware." Here we have expanded on those concepts, considering the overlap and spaces between hard, hot-work operational locales and soft, in-house fire station activities. Hard and soft environments include any and all activities that take place when firefighters are preparing, standing by, responding, operating, standing down, terminating, or returning; the soft environment is essentially the staging area for the hard environment. The pre- and postresponse modes of the fire, rescue, and emergency business must be augmented by *hard-environment mind-sets* and *soft-environment mind-sets*. Hot-work, in-action modalities and emergency situations always demand a hard-environment mind-set.

This textbook also introduces the concepts of *hard-think* versus *soft-think* attitudes. Hard-think is required when any activity, action, or behavior can, will, or may directly or indirectly affect hard-environment operations. Because all live and hot training operations are hard environments, training requires hard-think (fig. I–4).

Soft-think is an on-duty luxury and is warranted only when hard-think is not required. Hard-think is required when equipment is being stored, inspected, maintained, tested, inventoried, placed in service, or taken out of service. Hard-think is required whenever any apparatus' is ignition-on, regardless whether idling, moving, pulling out of quarters, backing into quarters, responding to alarms, or returning from any operation. Hard-think is required when apparatus are in the process of parking or in operation for any reason. Hard-think is required anytime units are on city streets—even during nonemergency use. Thus, hard-think is the "think of choice" when preparing, standing by, operating, standing down, or terminating activities that relate to life safety and property protection.

The attitudes and actions in the soft environment almost always affect hard-environment performance. If the soft environment stinks, there is little chance of a non-stinky hard environment.

The entire universe is made up of matter. In basic terms, matter is the stuff that can be seen and felt, and in the space between chunks of ordinary matter that can be seen and felt, there is a substance known as "dark matter." Hard-think is a bit like dark matter: You may not be able to see it, but hard-think is essential and should take up an awful lot of space in the fire, rescue, and emergency universe.

Fig. I–4. Company officers must ensure that hard-think attitudes permeate the training ground. Mistakes made here are more readily addressed than in the hard environment. (Courtesy Al Pratts)

Many excellent management and leadership texts are available to interested readers. These texts allow in-depth research into areas beyond the scope of this book, such as foundational management theory, leadership information, and command and supervisory tools and skills.

As Avillo sees it

Because there is a gray area between the hard and soft worlds, I offer this rule of thumb: any activity addressed in the soft environment that has potential to affect the hard environment in any way must be approached with hard-think. This includes all activities involving life safety, incident stabilization, and property protection.

The following activities demand hard-think. This holds true from the first on-duty minute until the last. Hard-think applies to but is in no way limited to:

- Ensuring that all personal protective equipment (PPE) is maintained and ready for response, as well as properly donned and secured when used
- Keeping apparatus and tools in the proper condition and ready for use (fig. I–5)

Fig. I–5. Success in the hard, hot-work environment is the by-product of a lot of hard-think in softer environments. (Courtesy Ron Jeffers)

- Ensuring, through regular discussion and training, that operational plans are in place, known, and understood
- Conducting defensive driving operations at all times

- Holding preplanning activities
- Conducting training exercises including skill development, assignment preparation, and review of standard operating procedures and policies
- Supervising all soft and hard activities as if the lives of the members depend on it

Hard-think protects the leader and those being led from becoming links in a chain of negligence. Chains of negligence are a series of unintended negative consequences resulting from a failure to properly supervise. Maintaining a hard-environment mind-set is a safeguard against all forms of mini-negligence that can and will jump-start a chain of negligence.

A hard-environment mind-set can exist in the soft environment, but soft-think can never exist in the hard environment. A leader's mind can never dress casually for the hard party; formal hard attire is required at all times. In-service and ready status requires right-headed thinking for the team, company, battalion, or platoon to meet the demands of the hard, hot-work environments.

Basic functions of a leader

- *Planning.* Recognizing and prioritizing things that need to be done; determining methods to accomplish the goals set by the organization.
- *Organizing.* Establishing and supporting the formal chain of command; assigning tasks that are defined by the organization.
- *Staffing.* Training and assigning personnel to positions that support the formal organization.
- *Directing.* Making decisions; assigning tasks through orders and instructions.
- *Coordinating.* Linking interrelated functions; creating a common thread of activity and support for achievement of the goals of the organization.
- *Reporting.* Keeping superiors and subordinates informed of all relevant department business.
- *Budgeting.* All activities that go with budgeting, including planning, accounting, and control.

chapter 1
CONGRATULATIONS ON YOUR PROMOTION! NOW WHAT?

The man who commands efficiently must have obeyed others in the past, and the man who obeys dutifully is worthy of someday being a commander.

—Marcus Tullius Cicero

Show me a man who cannot bother to do the little things and I will show you a man who cannot be trusted to do big things.

—Lawrence Dale Bell

In the moment that it takes to raise one hand and take your oath, you enter into a legal, personal, and moral contract. The oath allows *entré* into a world of great privilege and responsibility. It is a great gift to be allowed to lead and command women and men who have selflessly chosen to do society's most dangerous work under the most extreme conditions.

The gift of leadership comes with great responsibility: You are no longer responsible for your actions alone. You are charged with bottom-up and top-down responsibilities. You are responsible for your crew, the apparatus, all assigned equipment, and the facility in which your company is stationed. A fire service leader must demonstrate loyalty to their superiors and carry out all lawful orders and directions; must enforce department policies and procedures; must support the mission of the formal organization (the fire department); must extend every effort to achieve the goals of the department; and must be committed to excellence 24 hours a day, seven days a week, and 365 days a year (fig. 1–1).

Fig. 1–1. Congratulations on your promotion! Immerse thyself!
(Courtesy Audra Carter)

Leadership carries a lot of responsibility, and with responsibility comes accountability. Leaders must do the right thing, protect their people, and serve the community through fire, rescue, and emergency response. The engine that drives responsible leading is loyalty. (Loyalty must be considered an action word. There is no room for passive loyalty.) A fire service leader must be loyal to the community served and the department as well as those they command (figs. 1–2 through 1–5).

Figs. 1–2 through 1–5. You now own everything. (Courtesy Ron Jeffers)

The Formal and Informal Organizations

The "Prime Directive" of every leadership position is to bring everyone home safely. The next most important directive for commanders is to protect their people from the consequences of self-inflicted wounds in the soft environment (within either the formal or informal organization). The formal organization describes the mission statement and defines the goals and objectives of a fire department; in other words, the fire department is the formal organization. The formal organization provides the framework that guides how an officer should lead and perform the business of fire, emergency, and rescue mitigation in both the soft and the hard environments (fig. 1–6). General orders, policies, standardized operating procedures (SOPs), and training should be designed to support and direct leaders in the tasks assigned. The formal organization is a support mechanism and safe guard and safe haven. The formal organization identifies and arbitrates the next right thing to do.

Fig. 1–6. The formal organization provides rules of engagement, secret handshakes, and means and methods.

As an officer, you will be supported by the formal organization in direct proportion to the support you provide to the department. Playing outside formal organization boundaries throws everyone's game off. If you choose to play by your own rules and stray from the set parameters, then you must be prepared to accept responsibility for any and all consequences that result from your conduct. Loss of command and control increases risk for all personnel on the emergency ground. Confusion, uncertainty, inconsistency, and failures of leadership are the usual by-products of straying from the formal game plan (fig. 1–7).

As a representative of the formal organization, you must recognize all the forces at work, including the informal organization, which is an organization within the organization. The informal group dynamic is an alternative-reality mechanism that is at work 24 hours a day, seven days a week, 365 days a year. The informal organization thrives on gossip, rumors, scuttlebutt, and other types of unofficial information. In the informal organization, activities contrary to the mission of the organization can find root and grow. Nevertheless, the informal organization is not all evil, and when properly directed, it can complement the formal structure and become an invaluable asset.

Fig. 1–7. New firefighter recruits depend on the organization to guide, nurture, and support their career. Inconsistencies caused by straying from the formal game plan of the organization can jeopardize an entire career. (Courtesy Ron Jeffers)

Effective leaders recognize the existence and power of both the formal and the informal organizations. An effective leader will work with the informal group to get the best out of the job, themselves, and their subordinates. An officer must align the needs of the informal group with the goals and mission of the formal organization. Because the informal group is not going anywhere, a fire service leader must become adept at finding the intersection of *Informal Group Freeway* and *Formal Organization Boulevard*.

The best way to counter the negative effects that the informal group can inflict on the formal mission and your company, battalion, or division (your little piece of the world) is to ensure that proper information is available. Effective leaders must keep the channels of communication open, providing the people who report to them with information that is clear, current, correct, and digestible. Officers who refuse to acknowledge the informal group will undermine and dilute their own ability to lead. Informal groups are part and parcel of the fire business and this reality should be incorporated into leadership strategies.

Leadership in the Fire Service

Leadership is all about human interaction. To be an effective leader, you must become a student of human relations. Leadership involves:

- Establishing a clear vision
- Sharing that vision with others so that they will follow willingly
- Providing information and outlining methods that lead to the realization of that vision
- Balancing the conflicting interests of all members and stakeholders

A leader in the fire service needs to be many things. The following sections detail eight criteria that a creative, effective, and respected fire service leader must meet.

Be consistent

An employee-centered fire service leader must understand that consistency is the key to effective and successful leadership. A fire service leader must earn the trust of their subordinates and superiors. Consistency is about setting clear expectations and following through. It is about behaving in alignment with department values. It is also about keeping promises or renegotiating promises if you cannot keep them.

Consistency cannot exist in a vacuum. Honesty, emotional maturity, fairness, competency, dependability, organization, intellectual curiosity, and courage of conviction are all components of an effective leadership philosophy.

Be honest

Honesty and integrity are synonyms. Being honest with peers, subordinates, and superiors entails making a conscious decision. Honesty and integrity may seem to be easy choices; however, acting with integrity and being honest are not always the easiest path.

Being honest and consistent requires a leader to be straightforward in word and deed. Moreover, being honest often requires a leader to be confrontational. Nevertheless, it is always better to tell the hard truth than it is to try to keep the peace by being dishonest. Honesty keeps a leader and those being led out of trouble. It is always better to be honest and truthful.

Honesty reveals character. Character is the lens your co-workers view you through. When we are honest, we build credibility. Being credible simply means that our words line up with the truth. Dealing honestly with people and situations enables consistency of behavior. Honesty reinforces the perception of a leader's trustworthiness and loyalty to subordinates, superiors, and the organization. Conversely, dishonesty is a credibility killer.

Be emotionally mature

Emotional maturity is measured by how a leader deals with situations and controls emotions when dealing with others. A leader with emotional maturity accepts responsibility for personal feelings, experiences, behavior, and life circumstances. Emotional maturity refers to the ability to understand and manage emotions.

Emotional maturity is reflected by your thoughts and behaviors. When faced with a difficult situation, your level of emotional maturity is one of the biggest factors in determining your ability to cope. Adrenaline-driven decisions and actions have no place in any leader's skill portfolio.

Interestingly, emotional maturity is contagious. When an individual responds with emotional maturity, those they are interacting with often start exhibiting symptoms of emotional maturity. Emotional maturity is not determined exclusively by each individual; instead, the level of emotional maturity exhibited by an organization will affect the maturity levels of its firefighters, fire officers, and chief officers.

Be fair

Fairness is the quality of making judgments that are free from discrimination. Fire service leaders must strive to practice fairness in all dealings with subordinates and superiors alike. How a leader deals with violations and infractions contributes greatly to how that leader is perceived as a consistent leader. Leaders in the fire service must:

- Play by the rules
- Take turns and share
- Be open-minded
- Listen to others
- Never take advantage of others
- Never blame others carelessly
- Treat everyone fairly

Be competent

Competency in a job is not the same as competency at a task. Leadership competency relates to how a fire officer carries out the responsibilities of the rank and functions assigned. A leadership competency skill set covers personal and professional abilities. Competency requires a comprehensive understanding of department policies and procedures and the ability to plan and prioritize work projects. A competent fire officer can identify situations that are or may become problematic and decide on the action(s) needed to correct the situation. A competent fire service leader can work with and embrace new methods and technologies.

Professional competency skills can be measured by how leaders and their subordinates comply with department policies and procedures. A competent fire service leader must be able to communicate effectively in writing and orally.

Without question, every fire service leader must continually demonstrate competency in developing and supporting teams. A competent leader is willing to confer and collaborate with peers, subordinates, and superiors. Every fire service leader must demonstrate the capacity to build quality work relationships and communicate sincere interest for the group's well-being. A competent leader will look to utilize the strengths of others and give them opportunities to contribute to the mission, task, or assignment.

Be organized

Organization facilitates a leader's capacity for competence, consistency, and fairness. Effective leadership cannot coexist with the chaos of disorganization. Organizational skills begin with clarifying priorities and objectives. Decide what you want your command to look like. An effective leader will have a personal command image, for good or ill: Will yours be General Washington or General Wishy-Washington?

Organization does not just happen—it must be developed and practiced. Learn a system to process records, files, reports, and other paperwork. Deal with all administrative tasks in an ordered and professional manner.

Learn how to prioritize and schedule. Incorporate alternatives in case plan A goes awry. Organization frees a leader to focus on what is truly important by identifying what is unnecessary. Organization allows a leader to evaluate activities, routines, and systems that are not working. Recognizing what does not work allows leaders to focus on the activities that are necessary to achieve success.

Schedules and checklists are two of the simplest and most effective organizational tools available for line officer planning and execution (figs. 1–8 and 1–9). All assignments require some or all of the following:

- Time
- Staffing
- Funding
- Tools
- Equipment
- Resources
- Facilities
- Documentation

Without a plan, assembling any type or number of assets is a monumental waste of time. Action without planning is a form of Russian roulette: Spin the cylinder enough times, and the results will be disaster, tragedy, or death. Creating a checklist and a schedule eliminates the downside of Russian roulette–style planning.

Competent planning requires attention to detail. The failure to recognize and attend to details will invariably blow the best plans out of the water because each detail can have an exponential effect; therefore, there is no "little" detail. Regardless of the task, think it through and make sure you have the proper resources.

North Hudson Regional Fire & Rescue Monthly Activity Schedule				
Date:	Date:	Date:	Date:
Personnel not available for training	Personnel not available for training	Personnel not available for training	Personnel not available for training
Activities:	Activities:	Activities:	Activities:
Date:	Date:	Date:	Date:
Personnel not available for training	Personnel not available for training	Personnel not available for training	Personnel not available for training
Activities:	Activities:	Activities:	Activities:

Division _____ Battalion _____ Month / Year: _____

Battalion Commander _____

Fig. 1–8. Monthly training schedule

North Hudson Regional Fire & Rescue

DRILL SESSION SAFETY CHECK LIST

Building Address: _____

- [ ] Notify dispatch office.
- [ ] Set up command post; identify location to all.
- [ ] Establish communications frequency.
- [ ] Obtain required water supply.
- [ ] Position pumping apparatus.
- [ ] Position hoselines as needed for drill.
- [ ] Eliminate unnecessary debris inside and outside of building.
- [ ] Brief personnel on safety hazards; condition of building, hazardous, or unusual conditions.
- [ ] Assign ICS positions, instructors, monitors, and teams.
- [ ] Assign additional safety officers as needed.
- [ ] Establish emergency evacuation signal and demonstrate to all.
- [ ] Establish emergency escape routes from building and demonstrate to all.
- [ ] Establish emergency evacuation assembly area and demonstrate to all.
- [ ] Conduct pretraining briefing; explain objectives, scenario, procedures, restrictions, emergency procedures.
- [ ] Have students familiarize themselves with building layout, escape procedures, and routes.
- [ ] Establish emergency lighting.
- [ ] Raise ladders as needed for evacuation of roof/upper floors.
- [ ] Designate monitors inside building with thermal imaging cameras and hand lights; assign duties and responsibilities to monitors.
- [ ] Egress routes should be unlocked and unobstructed.

CRITIQUE

- [ ] All persons accounted for.
- [ ] Building inspected for stability and hazards.
- [ ] Overall training critique conducted.
- [ ] Records and reports prepared, as required:
- [ ] Accounting of activities conducted.
- [ ] Documentation of unusual conditions or events.
- [ ] If injuries occurred prepare reports and notify supervisor.
- [ ] Changes or deterioration of training center noted and reported.

Safety Officer: _____ Date: _____

Chief Officer: _____ Date: _____

Times of Training Session: _____ to: _____ Platoon _____

Companies at Drill: _____

Fig. 1–9. Drill safety checklist

Coordination with agencies, units, and teams is a nonnegotiable component of the planning process. Another nonnegotiable component of planning is having contingencies in case the mission goes bad or simply does not achieve the intended goals.

Competent leaders must be detail oriented. Competent, detail-oriented leaders make the most educated decisions. The devil and the genius will be found in the details. We offer the following quotes:

A schedule defends from chaos and whim. It is a net for catching days. It is scaffolding on which a worker can stand and labor with both hands at sections of time.

—Annie Dillard

Science is organized knowledge. Wisdom is organized life.

—Immanuel Kant

Don't agonize, organize.

—Nancy Pelosi

An idea can only become a reality once it is broken down into organized, actionable elements.

—Scott Belsky

Be intellectually curious

Intellectual curiosity reflects a leader's desire to invest time and energy in learning more about the "JOB" . Intellectual curiosity means having a genuine interest in knowing about and understanding the people, the organization, and the mission. An intellectually curious leader studies the roles and responsibilities at every level and position in the organization. The intellectually curious officer is a student of the science, arts, traditions, and history of the fire service.

Being intellectually curious means wanting to know more than just the basics and going beyond the knowledge base required to perform the "JOB" An intellectually curious leader wants knowledge that will be personally beneficial and help others and the organization.

Intellectually curious leaders seek to challenge themselves and those they work with. Intellectual curiosity creates an environment that makes people think about problems and issues. Intellectually curious leaders communicate with others about issues and find common ground to resolve issues and solve problems.

Be respectful of the power of orders

Failure to respect the power of orders will guarantee that your own orders will have little power. The way you follow orders is the way your orders will be followed. Flawed orders guarantee flawed results. The care and effort extended in carrying out an order will be equal to the care and effort invested in issuing an order. If you are not good at carrying out orders, then you will not be good at giving them. If an order cannot be enforced, then that order should not be given. Orders should not be sprayed around like cheap perfume.

Conclusion

To provide safe, effective, and positive leadership, a fire service leader must possess a comprehensive, knowledge-based understanding of the fire service. A leader must embrace and understand the mission, the structure, the history, the formal mandates, and the great traditions that are the foundation of the fire service (fig. 1–10).

Leaders realize that they do not know everything and will not always be right. Therefore, they must have the capacity to listen, accept new ideas, follow up, and conduct research instead of guessing. An effective leader must be familiar with the administrative mandates, tools, policies, procedures, and directives provided by the formal organization. A fire service leader recognizes that rules, regulations, and policies are guides designed to support safe and effective supervisory action and decision-making. The leader must respect, support, and understand the roles and responsibilities of every line, staff, and support position in the organization.

Fig. 1–10. Dedication plaque at North Hudson Regional Fire & Rescue headquarters—ensuring that the roots of tradition are not blurred by time

chapter 2
FULL-CONTACT LEADERSHIP

Leaders must be tough enough to fight, tender enough to cry, human enough to make mistakes, humble enough to admit them, strong enough to absorb the pain, and resilient enough to bounce back and keep on moving.

—Jesse Jackson Sr.

To create is to produce through artistic or imaginative effort.

—anonymous

In order. to seek one's own direction, one must simplify the mechanics of ordinary, everyday life.

—Plato

Introduction to Full-Contact Leadership

By the time I joined the fire department, I had been exposed to a variety of leadership styles. I had played sports, been a Boy Scout, done time in parochial and public school systems, explored higher education at a couple of colleges, done my bit for Uncle Sam, and worked a variety of jobs (not one of which could be called a career).

With experience as my only guide, I had developed skewed and poorly rooted beliefs about leadership. I thought being a leader meant lots of bright buttons and gold braid so I could lead the big parade. I truly believed my first name would be changed to "Sir," and I was sure that every maître d' would usher me directly to the best tables—in all the best fast food joints and diners in New Jersey! Not one of these leadership fairy tales ever came true. To the contrary, my buttons got tarnished, and the big parade often marched to a different drum. In 1983 I was promoted to the rank of captain and given command of a ladder company in the Weehawken, New Jersey, Fire Department.

It was only after being given the opportunity to lead a company of firefighters that I became aware of how little leadership has to do with a formal title. True leadership is a moral and legal, as well as formal and informal, contract between an officer and the operators assigned. A leader's position is strongest when subordinates, peers, and superiors recognize and acknowledge that the efforts, attitudes, and consistency of command behavior constitute always searching for a way to match the mission and goals of the unit, division, or department with the needs and aspirations of the men and women assigned (fig. 2–1). Recognition as a leader is hard won and forged in close quarters, through full-contact interaction with other humans.

Fig. 2–1. The goals of the officer (leader) must be balanced with and match those of the department.

chapter 2 | Full-Contact Leadership 17

This book speaks about full-contact leading. Full-contact leading means finding new, inventive uses for and combinations of tried and true leadership practices. Full-contact leadership is an individualized assemblage of human relationship skills, informed by life experience and committed, concentrated study of the management arts and sciences. An effective leader embraces the mission of the fire service and maintains a socialized orientation toward leading men and women in accomplishing the goals of the department. Full-contact leadership is a set of interpersonal skills, management theories, and principles brought together in situation-specific design, to achieve a goal, solve a problem, develop a subordinate, or support the formal organization (fig. 2–2).

Fig. 2–2. Situation-specific action planning. Goal achievement, problem-solving, subordinate development, and formal organization support provide fertile ground for full-contact leader development.

As used in this book, the term "full-contact leadership" describes a pathway to get the "JOB" done right by improvising with just about every leadership skill, term, theory, idea, philosophy, hypothesis, conjecture, action, procedure, policy, program, and so forth—the list goes on ad infinitum! Full-contact leadership is not a paint-by-numbers style of leadership. It is sophisticated, proactive, and requires an outside-the-box state of mind.

A full-contact leader does not dismiss any of the orthodox management theories, leadership skills or supervisory tools. Having knowledge of a subject does not automatically translate into having the capacity to effectively apply that information. A full-contact leader will aspire to move beyond just knowing something and will consciously search for ways to combine and apply the stuff in the books and put "that stuff" to work at street level (fig. 2–3).

Fig. 2–3. Full-contact leaders know the job, teach the job, and do the job. Consequently, they transform theories and principles into successful fire, rescue, and emergency operations.

Full-contact leadership is a service position. It serves the individual, the organization, and the mission in the hard and soft environments of the fire, rescue, and emergency mitigation business. At the foundation of the concept of full-contact leadership lies an understanding that one of the primary functions of leaders is to develop the people they are charged to lead.

People know leadership when they see it. Our experience and the people we have had the great good fortune to work with have taught us a new twist on an old cliché: You can fool all of the people some of the time, but you can fool the people you work with only once.

The reader is encouraged to put the information contained in this book to work in the streets. To accomplish this, leaders need to be students of the human condition. An effective leader must be imaginative and invent ways to ensure a supportive work environment. Full-contact leaders design a personalized, integrity-based leadership style that has positive, meaningful impact and moves the mission forward.

The Full-Contact Leader as Transformational Tool

Transformational. The act or an instance of transforming; the state of being transformed; a marked change, as in appearance or character, usually for the better.

Full-contact leaders have transformational power. They have a strong vision and strong convictions. Consistency, fairness, and positive attitudes provide subordinates, superiors, and peers with a role model who has the potential to change perceptions, motivate people, build teams, and transform work environments.

Challenging subordinates, creating situations that are intellectually stimulating, and providing individual consideration and attention can transform and motivate subordinates, peers, and superiors. Full-contact leaders encourage those they lead to come up with new and unique ways to challenge the status quo and to work within organizational pathways to find support for their projects and suggestions. Full-contact leaders encourage subordinates to think and perform in original and innovative ways; this is also known as "leading from below." Full-contact leaders are better equipped to find unique solutions to complicated problems because the full-contact leader is able to tap into all the talents and knowledge of the entire team.

When subordinates are allowed to hold greater ownership of their work, they tend to embrace changes more readily. Further, vital group image is enhanced, and self-esteem is improved. When work groups and the work being done are recognized, subordinates and superiors proudly buy into the collective identity of the organization. The formal organization must communicate that both the work being done and the workers doing the work are valued, respected, and appreciated.

The Full-Contact Leader as Servant-Leader

Stewardship is the careful and responsible management of something entrusted to one's care. A full-contact leader is both servant and steward. Fire service stewardship (leadership) is the act of accepting the assignment and assuming responsibility to safeguard, lead, supervise, and manage the personnel, equipment, apparatus, and facilities placed at one's command.

The term "servant-leader" was coined by management expert Robert K. Greenleaf in an essay published in 1970, "The Servant as Leader." In this essay, he wrote, "The servant-leader is servant first. . . . It begins with the natural feeling that one wants to serve, to serve first. Then conscious choice brings one to aspire to lead." Servant leadership is an ancient philosophy; there are even passages in the ancient text *The Art of War*, attributed to Sun Tzu around 500 BCE, that relate to servant leadership.

Servant leadership operates on a sliding scale between autocratic and laissez-faire leadership styles. Servant leadership finds the point of balance between leader and servant. Formal rank, authority, and leadership mandates are not abdicated in a servant leadership model. Position in the formal organization is not diminished by adopting a servant leadership philosophy.

Servant leadership should never be confused with weakness. Servant leadership is an exercise in courage, strength, and integrity. Thus, the servant-leader is a stewardship model for the fire service leader.

Never confuse servant leadership with the weak-kneed leadership, as shown in following example:

Firefighter Joe Schmoe arrives at the firehouse. Schmoe is met at the door by Captain D. Mat. Captain D. Mat holds out a steaming cup of cappuccino and inquires as to what tools and riding assignment Firefighter Schmoe would care to explore during his tour of duty. A little while later, Captain D. Mat offers to mop the floor and take out the trash.

This is an example of weak-kneed leadership. Servant-leadership never looks like this.

As Avillo Sees It: Weak-Kneed Leadership

Weak-kneed is defined as lacking strength of character or purpose or yielding readily to force, persuasion, or intimidation. A weak-kneed leader fails to embrace and defend the roles and responsibilities of command. Failure to embrace and defend these roles and responsibilities creates a command vacuum—and as with nature, command abhors a vacuum. The informal group will gladly fill any and every leadership void. When the leader abdicates to the informal group, that leader is signing up to become a victim, martyr, or manipulator.

A weak-kneed leader will be bullied by subordinates, peers, and superiors. Weak-kneed leaders become passive observers instead of active participants. No work gets done unless the crew wants to work. A company with a weak-kneed leader is crew driven and unsupervised.

Even when the negative consequences of failure to supervise become apparent, the weak-kneed leader is powerless to intervene.

To avoid confrontation, a weak-kneed leader will allow consequential dominoes to fall. The weak-kneed leader's response to situations gone bad is invariably regret. The weak-kneed leader rarely learns from the past and is not able to hold crew members accountable.

A weak-kneed leader is like an ostrich with its head in the sand. The leader is always hoping that problems go away by themselves and will often adopt an attitude along the lines of "Let's wait and see what happens next." This allows issues to run a deviant course without corrective intervention.

The weak-kneed leader is not unfamiliar or unaware of the roles and responsibilities of the position. The weak-kneed leader may care about the job and the personnel assigned but does not have the audacity to carry out the requirements of the formal organization.

A weak-kneed leader will focus on crew comfort instead of the demands of a given situation. Even if crew comfort is detrimental to the organization, crew comfort will trump the needs of the organization.

The weak-kneed leader is fearful, is afraid of superiors, and will make excuses for failing to meet the demands of the formal organization. The members of the crew come to expect the leader to cover for them. This seems kind of backward, right?!

The weak-kneed leader will overlook infractions if it means leaving his or her comfort zone or disturbing the comfort levels of team members. A weak-kneed leader is always hoping not to get caught by a superior; that is, getting away with it is considered a success. When department needs are in conflict with crew comfort, a weak-kneed leader is often reduced to pleading for compliance instead of ensuring compliance. A weak-kneed leader is too often dependent on coercive power, promises, rewards, and empty threats of negative discipline.

"Don't worry about it. Let's hope for a better outcome next time." "It's okay. It's no big deal." These are the sounds of weak-kneed leadership. Under weak-kneed leadership, the company, apparatus, equipment, and firehouse are unkempt; records are inaccurate or incomplete, and reports are vague. Meals and extradepartmental activities are the most important activities of the day, whereas department activities are looked upon as either a nuisance or a necessary evil. The weak-kneed leader knows that this is wrong but is powerless to change, hoping instead that no one will notice.

Weak-kneed leadership is a symptom of organizational malaise. Weak-kneed leadership can exist only if the department does not hold all levels of command accountable for any and all failures to supervise. The department must focus time, energy, and resources to ensure that all command positions are being filled by competent, capable, mission-focused, and employee-centered leaders (fig. 2–4). Weak-kneed leaders can exist only in a weak-kneed department.

chapter 2 | Full-Contact Leadership 23

Fig. 2–4. There is no room for weak-kneed leadership in this environment. Weak-kneed leadership must be recognized well before this situation rears its head. (Courtesy Ron Jeffers)

chapter 3

BASIC STYLES AND FUNCTIONS OF LEADERSHIP

Leaders must be close enough to relate to others, but far enough ahead to motivate them.

—John Maxwell

No leader should be totally dependent on any one single leadership style. No leader should disregard or dismiss any leadership style. Each of the four basic leadership styles described in this discussion has a place in every leadership skill portfolio.

The most creative leaders can recognize which leadership style is appropriate to a given situation. More often than not, any one, two, three, or all four styles can (and need to) be combined to create a recipe for success.

Autocratic Leadership

In a nutshell: *My way or the highway.*

Autocratic leadership is sometimes known as authoritarian leadership or dictatorial leadership. In this form of leadership, a single individual controls all decisions, decides the direction of the organization, and issues all orders. Autocratic leadership provides almost no opportunity for input from group members because such leaders rarely seek feedback or accept advice. Decision-making is instead based on the leader's own beliefs, ideas, judgments, perceptions, and experience. Autocratic leadership involves absolute authoritarian control over a group.

Autocratic leadership is best reserved for situations that are urgent, stressful, and time sensitive that require strong, decisive direction. Military combat, delicate surgical procedures, and fire, rescue, and emergency operations benefit from autocratic leadership. For example, jobs need to be accomplished quickly and efficiently on the fire ground, and autocratic leadership provides a highly efficient decision-making process. Consequently, this is the predominant leadership style used at fire, emergency, and rescue operations. Receiving unambiguous directions, fire operators can focus on assigned duties and the specific tasks at hand. Autocratic leadership relieves teams and team members of the need to make complex decisions during time-sensitive and urgent situations.

Autocratic leadership is not always issued through verbal commands. Response assignments, department rules and regulations, roll call, and off-duty times are examples of autocratic leadership in action (fig. 3–1). SOPs, personal protective equipment (PPE) requirements, the incident management system, and radio and fire ground protocols are written forms of autocratic leadership (fig. 3–2).

Fig. 3–1. Face-to-face verbal commands are autocratic leadership in action. (Courtesy Ron Jeffers)

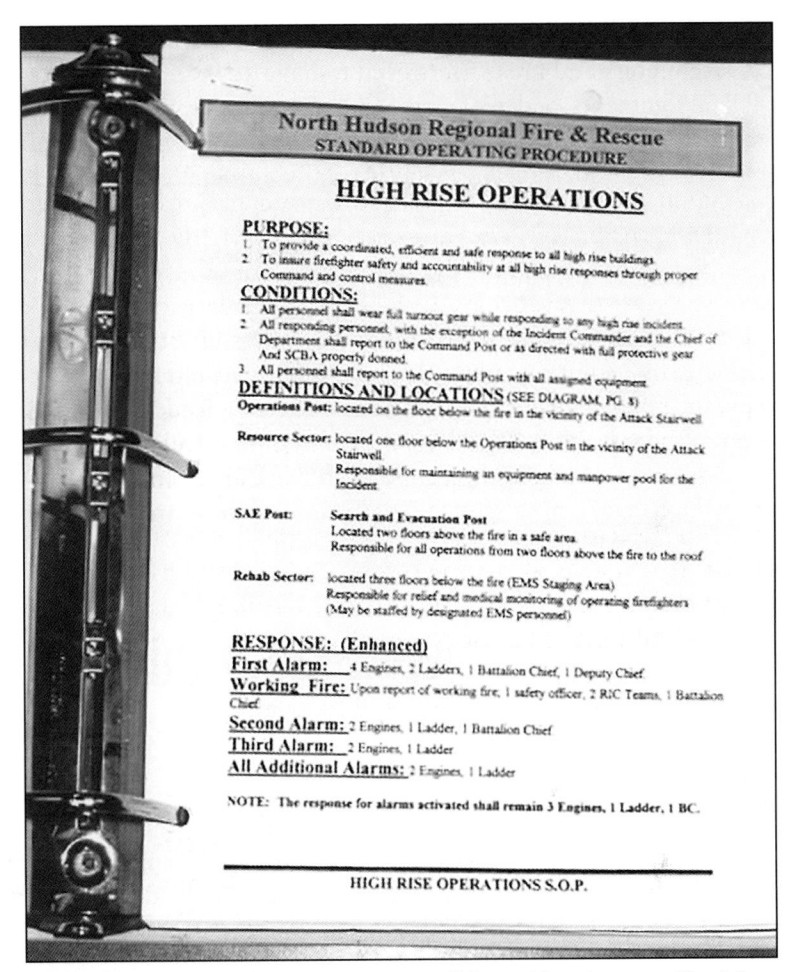

Fig. 3–2. Nothing a leader wants to accomplish would make sense if not for SOPs, rules and regulations, policies, and training.

A full-contact leader can access and integrate democratic and participative styles of leadership within the framework of autocratic leadership. Although autocratic leadership does have potential pitfalls, leaders can learn to use elements of this style wisely. For example, an autocratic style can be used effectively in situations where the leader is the most knowledgeable member of the group or has access to information that other members of the group do not.

Few would debate that autocratic leadership has a place in the fire service. Nevertheless, there are also many instances where this leadership style can be problematic. Leaders who use an authoritarian style all the time tend to abuse subordinates and foster resentment and resistance. Firefighters and fire officers need to have an investment and a say in their assignments and work environment. Autocratic leaders make decisions with little or no consultation with groups or group members. People working under a leader who is totally autocratic are denied any input, are allowed no opportunity for feedback, and have no personal investment in their work. Minimal subordinate development takes place when autocratic leadership is the only leadership style used.

Researchers have found that autocratic leadership often results in a lack of creative solutions to problems, which can ultimately hurt the performance of the group. Successful and effective leadership is dependent on a leader's ability to size up each situation. A full-contact leader will access and introduce the types and styles of leadership that will provide the safest, most effective path of least resistance required to achieve the intended goal.

Participative Leadership

In a nutshell: *Let's have input from everyone. Then we'll run it up the flagpole and see who salutes.*

Participative leadership is a style of leadership that encourages team members to work together to identify goals and then develop procedures and strategies to achieve those goals. This leadership style requires that the leader become a facilitator. Participative leadership allows team members to assign tasks, accept responsibility, and determine the direction of project completion. Participative leadership allows employees to work together without a superior issuing orders or making assignments. One of the main benefits of this style is that the process allows for the development of future leaders.

Participative leadership encourages active involvement on the part of everyone on the team (fig. 3–3). Team members are able to express their creativity and demonstrate abilities and talents that would not be apparent otherwise. The discovery of hidden talents will benefit any team and alert the organization that there are people within the team who should be provided with opportunities to further develop some skill or ability for future use.

Fig. 3–3. Participative leadership allows the leader to encourage and support cooperation from the team, and all heads participate in moving the team forward.

Laissez-Faire Leadership

In a nutshell: *Everybody here is a rocket scientist. Go out, work it out, check with the other geniuses when needed, and bring me back a rocket. By the way, I as the leader will act as referee and arbitrator. If you weren't the best at what you do, you wouldn't be here.*

Laissez-faire leadership is also known as "delegated leadership." Laissez-faire leadership is a hands-off style that allows experimentation, individual research, creativity, and outside-the-box thinking. Individual group members are allowed to make decisions in this style, which is best used in situations where group members are highly skilled, motivated experts capable of working on their area of responsibility without oversight. Because the group members are experts and have the knowledge and skills to work independently, they are capable of accomplishing tasks with little guidance.

Laissez-faire leadership implies complete autonomy, which enables group members to feel more satisfied with their work. This style is best employed in situations where team members are passionate and inherently motivated by their work. Frequently, projects require experts from multiple disciplines to work collegially. Under laissez-faire leadership, team members are encouraged (and often required) to confer with, contribute to and share insights, discoveries, and information with one another. An example of laissez-faire leadership at work would be the Manhattan Project, which brought together numerous great scientific minds to produce the first atomic bomb.

Democratic Leadership

In a nutshell: *Everybody has an equal say. One person, one vote.*

Democratic leadership gives everyone a seat at the table. In other words, every member of the team is allowed an equal say in the what, how, where, and when regarding a particular project or activity. Democratic discussions are relatively free flowing, are often redundant, and tend to go off course easily. This style requires that a leader facilitate the conversation, similar to a referee. Team members must be encouraged to share their ideas, synthesize information, and introduce practical, viable, and achievable goals.

A leader using a democratic approach must define, clarify, and communicate the final decisions and directions that will be taken to meet the group's goals. In a democratic leadership forum, a leader should put into writing the issues, decisions, and actions agreed on. A leader of a democratic process must create unity of purpose among all team members. It is not a sin for members to disagree even after a vote; however, it is a sin if the members do not agree to disagree.

Democratic leadership seeks to make decisions based on a group conscience. A leader using democratic leadership would be wise to keep in mind that "New Coke," the Edsel, green Jell-O, and the Titanic were all products of group conscience. Deciding what to have for lunch is a straightforward example of a case where democratic leadership can be applied.

The following characteristics embody effective democratic leadership:
- Keep communication open
- Focus the discussion
- Be ready to commit
- Respect the ideas
- Explain but do not apologize

Integrating Leadership Styles Allows the Basic Functions of a Leader to Be Accomplished

Planning: Recognizing and prioritizing what need to be done. Determining methods to accomplish the goals set by the organization.

Organizing: Establishing and supporting the formal chain of command. Assigning tasks that are defined by the organization.

Staffing: Training and assigning personnel to positions that support the formal organization.

Directing: Making decisions. Assigning tasks through orders and instructions.

Coordinating: Linking interrelated functions. Creating a common thread of activity and support for achievement of the goals of the organization.

Reporting: Keeping superiors and subordinates informed regarding all relevant department business.

Budgeting: All budgeting activities: planning, accounting, and control.

chapter 4

GROUP DYNAMICS AND MASLOW'S HIERARCHY OF NEEDS

Every leader should possess an understanding of social theories including group dynamics and Maslow's hierarchy of human needs, which can be used to gain insight into the factors underlying common behaviors found in the firehouse and other work venues. An effective leader needs tools to disassemble and reassemble the emotional give-and-take inherent in all the interactions between subordinates, peers, and superiors. Further, understanding the fundamental concepts behind these interactions can be applied to the day-to-day operations of the fire, rescue, and emergency mitigation businesses.

Being an officer in today's fire service requires one to be a career student of the art and sciences of working with, managing, and leading other human beings. This philosophy, which we support and encourage, is the price that must be paid for swearing an oath, and every leader (including aspiring leaders) must commit to it. Taking the oath for the rank and position of fire officer is a legal and moral contract between you and those assigned to your command.

This chapter is a *Reader's Digest*–type review of the theories of group dynamics and Maslow's hierarchy of needs. Interested readers are also encouraged to seek out and avail themselves of the many excellent management books that explore these social theories in depth.

Group Dynamics

The fire department is divided into two groups: the *formal* group (formal organization) and the *informal* group. Formal groups are defined by a fixed set of rules, policies, and procedures. The framework of a formal group is established in writing, with rules and regulations that leave little room for discretion or interpretation.

Informal groups, by contrast, are social structures that develop within the formal organization. Informal groups consist of a dynamic set of personal relationships, brought together by common interests.

To be effective, a leader needs to accept the reality and influence of informal groups. Harnessed correctly, the informal group complements and supports the goals of the formal organization. Fostering the informal group can promote innovation and enable people to solve problems that require collaboration across boundaries. Thus, the informal group can create an environment that cultivates self-discipline within the group.

Leadership Roles

The formal organization designates and supports a formal leader (fig. 4–1). The position of formal leader is static; however, the person filling that role may change. The formal leader exercises authority conferred by the formal organization. Formal leadership has four key concepts:

1. Formal leaders are appointed by the organization.
2. A formal leader's scope of power is defined by the organization's official rules, procedures, and institutionalized customs.
3. A formal leader's behavior is also constrained by these rules, procedures, and customs.
4. The formal leader is required to interact with outside agencies and individuals.

chapter 4 | Group Dynamics and Maslow's Hierarchy of Needs   35

Fig. 4–1. The formal leader represents the organization on all matters and is the face of the department. (Courtesy Ron Jeffers)

The informal leader is someone in an organization or work unit who, by virtue of others' perceptions, is recognized as worth paying attention to or following. The distinguishing factor between the two leadership roles is that the informal leader does not necessarily hold a position of power or formal authority over those who follow. The informal leader influences others by acting in ways that evoke respect, confidence, and trust. An informal leader often does not intentionally try to lead.

Informal leaders can be exceedingly valuable to organizations and to the success of formal leaders. On the other hand, when informal leaders do not support the formal leader's agendas and vision, they can function as barriers within the organization. There can be occasions when the formal leader is also the informal leader.

Maslow's Hierarchy of Needs

Abraham Maslow modeled human needs as a pyramid. The more fundamental needs form the base of this pyramid. Individuals generally do not try to fulfill higher needs until the lower needs are met. These are categorized from fundamental to higher needs as follows:

- *Biological needs.* Food, air, water, clothing, and shelter are the most basic needs, and thus are represented at the base of the pyramid. Only after meeting these needs will individuals try to fulfill the higher-level needs. The most common method of meeting these needs is through an occupation. Individuals work for pay, and the money earned is exchanged to meet these life-supporting needs.
- *Security.* This is insulation or protection from events that threaten the continuance of meeting biological needs. Security also entails maintaining and expanding the level of need attainment. Having a savings account or a pension plan along with Social Security and insurance policies is an example of security development.
- *Social needs.* These are met through human relationships. Social needs include the need for attention, the need to belong, and the need to be respected by peers. It is at this level that the basis for informal group behavior is formed. Peer pressure is how the informal group influences the direction and motivation of individuals in the informal group.
- *Self-esteem.* The needs of this level involve ego fulfillment through individual recognition, status, and achievement. When self-esteem needs are met, an individual's self-image is fortified, and that person's concept of personal worth is heightened.
- *Self-realization.* This is the highest level of need. Self-realization is fulfilled when an individual feels that the limits of his or her abilities have been reached, that creativity has been used in doing so, and that the tasks that needed to be accomplished have been accomplished.

Society, along with a firefighter's station in life, facilitates the fulfillment of basic, biological needs. Consequently, firefighters may focus their time and energy fulfilling higher-level (security, social, self-esteem, and self-realization) needs. Recognizing which level of need a person is trying to meet gives the fire service leader a metric for studying and assessing individual behaviors within groups. Most needs that firefighters work to achieve will be at the top levels. Effective leaders will apply these principles of social theory (group dynamics and Maslow's hierarchy of needs) to the fire service to understand and develop the personnel they command.

chapter 5
RANK HAS ITS PRIVILEGES

Cast a long shadow.

—Battalion Chief Steve Quidor

Leadership is doing what is right when no one is watching.

—George Van Valkenburg

Your role as a leader is even more important than you might imagine. You have the power to help people become winners.

—Ken Blanchard

A leader without followers is just a man taking a walk.

—A. Lincoln

Leadership is a privilege to better the lives of others. It is not an opportunity to satisfy personal greed.

—Mwai Kibaki

Rank does not confer privilege or give power. It imposes responsibility.

—Peter Drucker

The word knight, which originally meant boy or servant, was particularly applied to a young man after he was admitted to the privilege of bearing arms.

—Thomas Bulfinch

For over thousands of years Roman conquerors returning from war enjoyed the honor of a triumph—a tumultuous parade. In the procession came trumpeters and musicians and strange animals from the conquered territories with carts laden with treasure and armament. The conqueror rode in a triumphal chariot, the dazed prisoners walking in chains ahead of him. Sometimes his children, robed in all white stood with him in the chariot or rode along on trace horses. A slave stood behind the conqueror holding a golden crown, whispering in his ear the warning: that all glory is fleeting.

—George Patton

"Rank has its privileges" (RHIP) is a familiar expression. RHIP asserts that higher ranks should have greater advantages or more rights than their subordinates. Even though this is a great philosophy for those who hold rank, there is little reason to believe that it is espoused by rank-and-file firefighters. The RHIP concept was undoubtedly hatched during a clandestine, secret-handshake officers' meeting way back when dinosaurs roamed the fire service world.

Fire service leader is a privileged position. The privilege is not about prime parking spots, primo vacation picks, or any other perks. Rather, a leader's great privilege is being entrusted with the safety, well-being, and development of men and women assigned to their command. The privilege of leadership includes earning the respect of those the leader serves, which entails living up to a certain standard and treating people with respect while effectively and professionally meeting the goals of the formal organization (fig. 5–1).

Leadership is not a matter of promotions or orders. A leader is a person who values others and has the judgment to do right by them. Leading high-functioning men and women who do dangerous work in close-knit units is a challenge. Leadership is often practiced on the steep side of a slippery slope! Nevertheless, leaders embrace this challenge, and when one challenge has been met, a full-contact leader moves on to the next.

Fire service leaders are not intended to make life easy for subordinates. The best a leader can do is not make life unnecessarily hard. The unjustified belief that no one should be uncomfortable is a dark alley that many would-be leaders get mugged in. Most, if not all, of life's rewards are derived when operating outside one's comfort zones.

chapter 5 | Rank Has Its Privileges 39

Fig. 5–1. Rank does have privilege, but responsibility outweighs privilege every time. A full-contact leader talks the talk, walks the walk, walks the talk, and talks the walk. (Courtesy Ron Jeffers)

The philosophy of the very first company officer I worked for was, "I was wrong once, but that turned out to be a mistake." Being a leader should never be confused with being an expert. A leader must know when to allow someone else to point the way. Achieving a goal by way of the safest, most effective path of least resistance is what leadership is about.

Much will be revealed, and a lot of things will fall naturally into place when a leader's focus is on teaching, coaching, supporting, and developing. Any leader not teaching, coaching, supporting, or developing should reassess, reevaluate, self-evaluate, make changes, or move along. The positive impact of a motivated, educated, creative, and intellectually courageous fire officer can do more to advance an organization's mission than any other available asset.

Fire officers are "stuff magnets." People come to their officer because they need stuff, want to report stuff, find out about stuff, return stuff, complain about stuff, get you to fix stuff, change your mind about stuff, start stuff, stop stuff, etcetera and ad infinitum. Leaders, officers, and chief officers must accept that "stuff-getting" and problem-solving constitute business as usual. At the same time, a fire service officer needs to be judicious about when, where, and how far to stick their nose into the tent. Not all problems are problematic, some stuff cannot be got, not all feelings are facts, and there is no way to please everyone. Regardless how much or little gravity a leader assigns to an issue, remember that the problem may not always be important, but the person who brings forward the issue is always important.

It is not possible, necessary, or advisable for a leader to purge a team of problem-finders, complainers, or gripers. Problems, complaints, gripes, and so forth are how groups and individuals off-gas. Problems and complaints are points of contact between the formal leader and the formal and informal groups. A creative and effective full-contact leader will recognize "off-gassing" behaviors as opportunities to move the group or individual in a positive direction.

If complaining, moaning, groaning, and griping are out of proportion with overall conditions, then there is surely some issue to be identified and addressed, even if it is not yet on the radar. Working below the radar is where full-contact leaders can make the greatest contribution to the group and the organization. An effective leader will recognize that one or more of Maslow's levels of needs are being asked to be met.

Full-contact leaders respect each individual and every issue brought to their attention. However, there are certainly minor and insignificant issues. A leader must never diminish the stature of a subordinate bringing a minor or insignificant issue forward.

Some issues, concerns, and problems require more urgent attention than others. Prioritizing and dealing with issues requires leaders to triage any and all concerns that demand their attention. An issue that has any impact on safety of personnel, life safety, or property conservation is the most urgent—priority one—and must be dealt with appropriately and immediately. Whether in a soft or hard environment, a leader should never hesitate to act when life-safety or property conservation issues are brought to their attention (fig. 5–2). Any such concern should never be relegated to the "I'll take care of it after I finish my coffee" bin.

Given that problems are pervasive, solutions must be accessible to the leader. Solving a commodity issue is fairly straightforward—either you procure or you do not. The stickier problems are those where the solution cannot be hand delivered.

chapter 5 | Rank Has Its Privileges 41

Fig. 5–2. Full-contact leaders address problems before they become tragedies. (Courtesy Ron Jeffers)

A full-contact leader must deal in solutions. A test of effective leadership is recognizing an issue before it becomes a problem or emergency. When a problem is identified, a full-contact leader will take the appropriate remedial action.

For a leader to stay in the solution, a change of perception may be required. Problems need to be recognized as assignments and opportunities. This slight change of perception removes the negative connotation associated with problems and thus allows the resolution process to flow from a positive mind-set. When there are only assignments and opportunities, there can be no problems.

The well-worn iceberg metaphor states that 20% of an iceberg is visible while the remaining 80% is below the surface. For leaders to truly resolve problems, the iceberg must always be flipped over. This flipping is accomplished through supervision. To supervise literally means having extra vision. To be effective, a leader must be present, engaged, and hypervigilant. Full-contact leaders must see above and below the surface to understand and contend with root issues. Leadership concepts, tools, and skills are the lenses through which a leader can view the supervisory landscape with clarity.

Officers must recognize that the formal organization does not hand out collar jewelry to impress friends and relatives. Officers, leaders, and chiefs are positioned in the chain of command to fix things. A leader wants to intervene before an issue becomes a problem. Solution-centered, proactive leadership is full-contact leadership at its best. Fire officers are like administrative, operational, and personality-based traffic cops. Fire officers clear jams, guiding and directing department business into the right lanes. They are strategically positioned to point out road signs and help subordinates avoid administrative potholes and disciplinary collisions.

Officers are tasked to resolve issues, solve problems, and deal with complaints as far down the chain of command or as close to the level of the problem as possible. Keeping the problem as far away from the chief of department as possible is a simple way to look at it. However, the capacity for resolving an issue is a function of the authority and responsibilities assigned. A company officer does not have the authority to declare "National No-Inspection Month," and neither can chief officers declare a "Paperwork Amnesty Day."

There are occasions when resolution of an issue requires channeling up the chain of command. Officers must be able to recognize when an issue requires the attention of a higher authority. A leader pushing an issue up the chain must provide a complete picture including the entire history behind the complaint. The officer must inform superiors of all actions taken to date and exactly where the issue stands presently. Some issues may require formal written reports, whereas others may be handled more effectively verbally.

A leader who brushes off a problem is brushing off the person who brought it up. A leader must keep in mind that bringing problems forth always has a personal component. A need to be accepted and respected is implicit in every exchange between leader and subordinate. When a leader disregards an issue brought forth by a subordinate, that sends a message dissuading others from coming forward. If a leader does not dignify small issues with due attention, then the group will be less willing to report larger problems. Conversely, leaders who invest appropriate time and attention to smaller issues receive dividends regarding larger issues.

Leaders are continually monitored and assessed by the people they lead. Every action and interaction is graded, catalogued, and filed. A leader's behavior conditions the behavior of the group and individuals being led. When a leader is consistent, engaged, and fair, the group will respond positively. By contrast, inconsistency will create unnecessary stress and fear, and lack of leader engagement will disenfranchise individuals.

To be successful, a leader must communicate in a consistently direct and frank manner. By contrast, indirect communication creates confusion, stress, and uncertainty, which undermines the health of the work environment.

Respect and common courtesy are integral to successful leadership. In the words of Deputy Chief Bob Montagne, "The bigger the man the bigger the courtesy." "The boss is tough but the boss is fair" may be the highest compliment any fire service leader can receive.

chapter 6

THE DICHOTOMY OF COMFORT

The comfort zone is the great enemy to creativity; moving beyond it necessitates intuition, которые in turn configures new perspectives and conquers fears.

—anonymous

The ultimate measure of an individual is not where he stands in moments of comfort, but where he stands in times of challenge and controversy.

—Martin Luther King Jr.

Prologue

Few professions are more rewarding and even fewer avocations more challenging than the fire service. The fire company is the cornerstone of every fire department. Every fire extinguished, every rescue ever made, and every emergency ever mitigated has been done by teams of men and women formed into engine, ladder, squad, and rescue companies.

Firefighters develop close relationships on and off duty (fig. 6–1). Camaraderie, mutual respect and support, and sister- and brotherhood develop through cooking and eating together, joking, and telling war stories; these interpersonal behaviors (the stuff they do together) forge powerful connections that make fire ground operators the most effective life-safety and community protection force imaginable. Firefighting, rescue operations, property protection, and emergency mitigation are activities performed every hour of every day by close-knit, professional, and familial units composed of the courageous, strong-minded, alpha-type individuals who are collectively known as fire companies.

Fig. 6–1. The second family. Those who function in situations of equal peril develop camaraderie and commitment to cause. (Courtesy Ron Jeffers)

Firefighters are human beings who do great service to society and the communities they serve. Close contact and complex relationships create favorable conditions and opportunity for conflict and controversy, which are by-products of humans interacting with other humans. For good and bad, fire companies can be hotbeds of controversy and conflict.

Conflict, controversy, disagreement, arguing, adult temper tantrums, moodiness, jealousy, and bullying are all part of the pantheon of misunderstandings, emotional toe-stubbings and ego collisions that are familiar, normal, and unsurprising conditions and behaviors that play out in firehouses across the fire service landscape. Because these behaviors are essentially normal and thus cannot be eliminated, officers should never be surprised when stuff happens. Thus, fire officers must possess and strive to perfect situational discrimination, high-level communication skills, appropriate attitudes, and the capacity and willingness to intervene, confronting and resolving conflict with maturity and efficiency.

As Avillo Sees It

Where do you stand in times of challenge and controversy? This question confronts fire service leaders every day. When a traffic light is green, when everything is running smooth, it is easy to stand up and be counted. Where and how you choose to stand when smooth turns into uncomfortable and confrontational is a good measure of the kind of leader you are. Creating and maintaining a healthy, productive work environment is one of the main functions of an officer, and this often requires the leader to be uncomfortable, interventionist, and confrontational.

Command-level comfort is an interesting concept. Supervisory failures and the inability to carry out assigned duties are often the results when officers are not able to move outside their individual comfort zones.

Fire service leaders need to take care of business. This is especially true when the show is not going as planned. When the integrity of the command environment is threatened, when conditions and situations are less than ideal, and when it is third down and long, a leader and the organization must step up, step in, and take care of business. Anyone can lead when conditions are perfect; situations that are less than ideal cannot be excuses for failure.

"Less than ideal" is exactly what leaders get paid to deal with. "Less than perfect" is an opportunity to improve, improvise, and do whatever is best. That is always a leader's best option. Doing something you think is right but having it blow up in your face is better than doing nothing at all. Trying and failing can be a learning experience. Doing nothing when something needs to be done is never an option. Conversely, doing something when nothing needs to be done—even if tempting, for personal gain or to enhance one's professional image—must be avoided.

A fire officer does not need to micromanage every situation. A full-contact leader can discriminate between situations that require intervention, those that require monitoring, and those that require a hands-off posture.

As we move up the promotional ladder, the organization rewards us with bars, horns, bugles, white helmets, gold buttons, cars, and offices where we can contemplate our navels. Formal adornments are not just indications of rank. Collar bling represents the responsibility and increased accountability that come with rank and command. Gold horns and silver bars require officers to be at their best when conditions are at their worst (fig. 6–2). Officers must prove the merit of their bars and bugles every single day in every single situation.

Fig. 6–2. Going to be a long night. It is in less than comfortable conditions that full-contact leadership pays dividends. Leaders do not have a choice; they must be at their best when conditions are at their worst. (Courtesy Ron Jeffers)

The quotation from Martin Luther King Jr. at the beginning of this chapter is about challenge and controversy, which in turn are synonymous with conflict and confrontation. Chief Flood talked often on the subject of conflict and confrontation. He held that there were four states of being in charge:

- Conflict
- Confrontation
- Resolution
- Confidence

chapter 6 | The Dichotomy of Comfort

For most people, conflict, confrontation, and resolution tend to cause discomfort. Nevertheless, none of these states is inherently bad. Simply put, conflict is a state of disagreement or disharmony between persons or ideas; confrontation happens when people encounter each other; and resolution is a course of action decided and agreed upon. People disagree all the time when they come into contact with other people. A leader is responsible for bringing people to agreement and creating harmony in the work environment. Leaders resolve things all the time, and they do this by deciding on courses of action. That is what confrontation, conflict, and resolution look like.

I learned to embrace conflict and confrontation as normal and natural functions of the human condition. I also accepted that conflict and confrontation were constant, inevitable, and just part of the job. Previously, failing to recognize the nature of conflict and confrontation invariably placed me at a disadvantage when dealing with the 'day-to-day' of leading the people in my charge. I came to see that conflict and confrontation were neutral zones where power was derived from the attitudes and actions of the parties engaged. The more maturely I insinuated myself into confrontational situations, the quicker I would be able to guide the conflict to resolution. Sometimes the path was smooth; other times the path was bumpy. Sometimes I could deal with situations informally; other times I needed to take formal action. Whatever the pathway, I was always able to find a way to resolve the issue, repair the condition, or address the situation.

My success at conflict resolution seemed pretty good, because I had always aimed to find compromise. Then I heard Chief Flood explaining that no one should begin at compromise. Because both parties have to give something up in a compromise, it is essentially a lose-lose situation. His suggestion was instead to look first for a creative solution that could be a win-win situation. If such a solution could not be found, then you could always fall back on compromise.

According to Chief Flood, the fourth state of being in charge is confidence. Confronting conflict in a direct and mature fashion while working to bring about resolution requires great personal courage and a subjugation of the leader's ego. Dealing with conflict, confrontation, and resolution builds layers of leadership confidence, which get thicker whenever leaving one's comfort zone. In other words, you must leave your ego at home when you leave your comfort zone.

Of course, there is a caveat. I was not always going to get it right. I was going to stick my proboscis where it did not always belong. In that case, I would make amends, step back, recalibrate, and learn from the experience.

Chief Flood told me that I would catch myself overlooking things that should have been addressed. In such cases, I should step in and take care of business as soon as I recognized that my attention was required. Proactive leadership is the ideal. Only when it's too late for proactive leadership is it time for reactive leadership.

On occasion an officer will find out about things when it seems too late. However, it is never really too late. If no formal action can be taken, the officer should inform the parties involved that he or she is now aware of what took place. In such a case, expectations need to be reinforced. The parties involved should consider themselves formally warned and be advised any future infractions will be subject to appropriate disciplinary response.

If you are not facing conflict and confrontation in the soft environment, you will not be effective in the hard environment. An officer must always keep in mind that the soft environment is just the staging area for subsequent action in the hard environment.

How conflicts, confrontation, and resolution are handled in the soft environment is an accurate barometer for how skillfully a leader can be expected to deal with situations in the hard environment. Leading in the face of challenge determines how a leader is perceived and respected by subordinates, peers, and superiors. This is true about challenges in both hard and soft environments.

It is appropriate to strive to be comfortable in your command. If a leader is not comfortable, that leader will be stressed and anxious. Because confidence leads to comfort, a leader must always work to improve all leadership abilities along with skill and knowledge portfolios. A leader must hold and maintain an edge, and confidence in your abilities is where you will find that edge. Seeking comfort in a command position is not the same as seeking confidence in a leadership role; comfort should be a function of confidence.

Here is where dichotomy enters the picture. A leader can be confident only when comfortable with being uncomfortable. A leader can be comfortable only after gaining the confidence to embrace and recognize uncomfortable situations as a routine part of the job.

An interventionist leader must maintain a delicate balance. Not every conflict requires the involvement of a leader. The men and women of the fire service are intelligent, motivated, mature, and mission-committed adults. Every officer must respect the individual and the informal group. The boundaries between formal and informal spheres of influence can be blurry. A leader's vigilance must be attuned to where, when, why, who, how, and mostly whether involvement is warranted. That is, a leader should exhibit intervention discretion.

Maintaining balance between intervention and abdication is a conundrum that every leader must contend with. This balancing act requires continuous attention and care. Supervision demands hyperawareness and vigilance regarding what is happening and what is going on.

> *Intervention discretion.* Just because you see it doesn't mean you need to react to it. But if you see it and it needs intervention, you better intervene.
>
> —Avillo

A leader must always be watching for issues and conditions that threaten the in-service and ready status of their command. Any condition, issue, action, or inaction that impairs the in-service and ready status of a command must be identified and then resolved. If subordinates know that their leader will never stick his or her head in the sand, they will never be surprised when the leader intervenes in conflict and confrontation situations.

Back in medieval days of kings and queens and guillotines, loyal subjects would do whatever they could to distract the king. The idea behind this was to make the royalty comfortable and keep them preoccupied. The king would be pampered, given exotic gifts, given excessive drink, and fed gluttonous amounts of food. By distracting the king, the subjects empowered themselves to rule the kingdom from below. Eventually the king would be assassinated—not a good thing if you are a king. Therefore, when an officer is much too comfortable, consider that officer essentially to be on the way to the guillotine.

A king should be wary of being too comfortable. Thus, a king would be wise to look beyond his comfort zone and beware of feeling overly comfortable in command. In other words, a king who removes his butt from the throne every once in a while, who gets his hands dirty, who takes a chance at being uncomfortable, and who addresses what needs addressing will live long enough to put his butt back on the throne.

> *The longer it takes a king to address issues that require attention, the larger the gap between acceptable leadership and unacceptable behavior and the shorter the reign.*
> —Avillo

Often, behaviors and actions of subordinates cause a leader to wonder if intervention is required. A simple size-up scheme can help answer the question of whether to get involved:

- Will anyone get hurt?
- Will anything get broken?
- Are any rules being broken?
- Will the unit's operational status be impaired?

Leader intervention is required if the answer to any of these questions is yes. Officers must intervene whenever they witness or have knowledge of actions, behaviors, or conditions that threaten life, cause injury, damage equipment, or are in violation of department rules, regulations, policies, or procedures. The action must be stopped, and the operators need to be educated, sent to neutral corners, or disciplined.

What if the operators work for another officer? What about unity of command? Well, the answer is simple: *You have to intervene anyway.* Unity of command is an organizational principle meaning one person, one boss. Unity of command is a foundational concept necessary to the efficient and proper operation of a formal organization. However, unity of command is not an invisible shield that allows an officer to disregard inappropriate actions or infractions, committed in their presence, by subordinates who are under the command of someone else.

Confronting a subordinate who is assigned to another officer is an uncomfortable dance. No officer likes it when another officer enforces a rule on or scolds one of their subordinates. This kind of situation reads very high on the discomfort meter!

All officers should respect unity of command. All officers are required to represent the formal organization (the department) whenever they witness or have knowledge of actions that are not in line with rules, regulations, procedures, or policies. Officers need to follow protocol when dealing with issues that include members not under their command. If the failure to comply affects safety, property, or equipment, a leader must address the issue immediately. Immediate action is expected when disgrace or dishonor could be brought to the department. Issues that do not affect life safety or cause damage to property or equipment and behaviors that can harm the reputation of the department should be referred to the individual's supervisor as soon as possible.

Few, if any, officers enjoy being told that one of their firefighters has screwed up or failed to comply. It is never comfortable to bring such information to the attention of a fellow officer. An officer's comfort zone should never trump a leader's responsibility to represent and be loyal to the formal organization.

Dealing with the behavior or actions of another officer's subordinate should be approached respectfully and understood to be officer-to-officer support. Supporting another officer, even when conditions are less than comfortable, is a proper and a positive form of representation for the formal organization.

Without exaggeration, it can be said that the fire service deals with real-world, life-and-death situations. Every misfeasance, nonfeasance, and malfeasance has the potential to be compounded exponentially, leading to tragic consequences. There can be no such thing as a squealer, snitch, rat, or stool pigeon. An officer's job is to protect the "JOB"—and thereby protect the operators doing the "JOB."

It is not our intention here to promote or legitimize vigilante leadership. Officers must acknowledge and support fellow leaders and their individual commands. Intervention must be tempered with respect and attention to proper protocol and intraunit integrity.

> *There is no such thing as an invisible shield of, "That firefighter doesn't work for me." An officer is an officer. An infraction is an infraction. This is one of the super-uncomfortable zones, so deal with it! Remember an officer is never "just passing through."*
>
> *—Avillo*

Leaders must be loyal to the formal organization, as well as to their subordinates. When subordinates are being treated unfairly by the formal organization, they should be confident that the leader will do the right thing and stand with and by them. Promoting a member to a leadership position is an implicit demonstration of trust in that leader. In the real world, there will be times when the formal organization is in the wrong; there will also be times when subordinates are in the wrong. Standing up to the formal organization when a subordinate is being treated unfairly is full-contact leadership at its best.

A leader must understand what it means to be a boss. Those being led must be educated regarding the responsibilities and authority assigned to a leadership position. All leaders must establish and clearly state their expectations.

> *I can be your boss or I can be your friend. If I have to choose, I will always be your boss.*
> —Battalion Chief Frank Vasta

> *Even a good friend can get you killed, but a good boss will get you home.*
> —Battalion Chief Steve Quidor

Officers should not surprise subordinates by improvising expectations based on the circumstances. Officers should never create an environment where subordinates need to guess whether the officer is friend or boss. Superior-subordinate boundaries should be understood and respected by both the officer and the subordinate. Setting expectations and defining boundaries demonstrates respect for subordinates and an understanding of the roles and responsibilities of the rank.

Firefighters and fire officers live together, support each other, and rely on each other. Firehouse culture is very social, often familial. Hence, formal responsibility and informal group behavior commonly become entangled. This entanglement has benefits but also makes for dangerous pitfalls. It is the responsibility of the officer to maintain propriety and disallow behavior that threatens the integrity of the command.

Sometimes in-house social issues can distract a leader. When social or informal issues conflict with the requirements of the department, an officer needs to remember where duty and loyalty lie, because otherwise the needs of the department suffer. If subordinates are conducting themselves in an unacceptable manner, it is the boss's job to straighten them out. Most wounds suffered by the informal group are self-inflicted. Very often, officer intervention saves the operators from themselves.

Every once in a while, one of my officers would be unwilling or unable to redirect his people onto the straight and narrow path. In such a case, that officer would receive my personal attention, guidance, and direction regarding the roles and responsibilities of a fire service leader. How far a leader allows his subordinates to stray from the formal organization is directly proportional to the time, energy, and grief that leader will expend in herding lost sheep back to the fold.

Recognizing an issue before it becomes a problem is a skill that defines full-contact leadership. When appropriate doses of tough love are administered, much pain and wear and tear can be avoided.

An officer is not one of the gang. A leader needs to create fraternization-free zones during on-duty time in the fire station. Thus, an officer should create a workday schedule that includes time away from the informal group. In my experience, any officer who is spending all day in the kitchen is not attending to very much department business. An effective leader knows when to separate himself from the herd.

chapter 7

POWER

Pull the string, and it will follow wherever you wish. Push it, and it will go nowhere at all.

—Dwight David Eisenhower

"Pushing on a string" is a figure of speech describing influence that is more effective in moving things in one direction than another. That is, you can *pull*, but not *push*, on a string. For example, if something is connected to you by a string, you can move it toward you by pulling on the string, but you can't move it in any direction by pushing on it.

Power is a string. Power should be recognized as a nonrenewable resource. As such, power should never be exerted without premeditation. Although it often is, power should not be placed in the hands of dilettantes, panderers, or pretenders.

Power may be the most misused and least understood of all the tools and skills available to a leader. Indiscriminate and careless application of power will undermine the authority of a leader. Reckless and uninformed use of power resonates throughout an organization. Incautious use of power can and will sabotage the goals and mission of the organization.

Uninformed and cavalier exploitation of power creates an environment of anxiety, resentment, and resistance. A leader wielding power in such an inappropriate manner will be embroiled in guerilla warfare with those being led. Subversion, distrust, blatant noncompliance, and surreptitious efforts to undermine authority are all predictable by-products of ill-used power. The stress, anxiety, and insecurity that result from abuse of power will be disruptive to the organization's mission.

When misused, power can metastasize through an entire organization like a cancer. The time, energy, and financial drain associated with undoing the damage associated with abuse of power are substantial and irretrievable. Thus, power must be applied with clear purpose. Power used judiciously, intelligently, and with compassion, can positively affect every strata of an organization.

Personalized versus Socialized Power Orientation

Nearly all men can stand adversity, but if you want to test a man's character, give him power.

—Abraham Lincoln

The personal desire to lead others is often powerful and compelling. Becoming an effective fire service leader and producing positive long-term results depends on the motivation that drives the desire to lead others. The desire and aspiration to lead others and seek power can be positive or negative depending on the motivation.

Two types of motivation indicate the intention of those who seek power. These are referred to as *personalized power orientation* and *socialized power orientation* These can be understood as a spectrum, because most fire officers' relationship with power will fall somewhere in between, rather than representing exclusively one or the other orientation.

Personalized power orientation

The measure of a man is what he does with power.

—Plato

Leaders with a personalized power orientation seek to gain power in order to aggrandize or glorify themselves and satisfy a strong need to fortify their self-esteem and status. A fire officer with a personalized power motivation will tend to exercise power impulsively and demonstrate little empathy, understanding, patience, or self-control.

Personalized leaders seek to dominate others. The personalized leader will use bullying tactics intended to keep subordinates weak and dependent. Personalized leaders need to collect and will always seek symbols of power and prestige. All authority to make important decisions is centered on the personally oriented leader.

Personally oriented leaders use coercive power to control and manipulate others. Individuals with a personalized power orientation lack empathy. They are self-centered, not very compassionate, often abrupt, and demanding. Leaders with personalized motivation are often insecure and rarely interested in debate, input, or feedback. The fire service leader with personalized motivation will respond to many questions regarding orders given with some version of "Because I said so!"

Departments, divisions, battalions, or companies that are under the command of a leader whose orientation is personalized will have a choked-off and restricted communication flow. Those being led will be slow to demonstrate initiative or to solve problems independently. Instead, subordinates will always be in standby mode, waiting for direct instructions from the leader. Scarcely any subordinate development will occur under a leader with personalized motivation. Subordinate enterprise and initiative will usually be suffocated. Command officers who are personally oriented will be hypervigilant and overprotective against any infringement on their authority. As the result of coercion, any subordinate loyalty will be situational and transient, possibly even feigned.

Socialized power orientation

Power isn't control at all—power is strength, and giving that strength to others. A leader isn't someone who forces others to make him stronger; a leader is someone willing to give his strength to others that they may have the strength to stand on their own.

—Beth Revis

Example is not the main thing in influencing others. It is the only thing.

—Albert Schweitzer

I cannot trust a man to control others who cannot control himself.

— Robert E. Lee

Leaders with a socialized power orientation understand that power must be used to the benefit of others. Being humans and firefighters, fire service leaders will by no means be perfect; however, leaders with socialized motivation are far less egotistical, defensive, and materialistic than the leaders with personalized motivation. Leaders with socialized motivation are generally more emotionally mature, empathetic, and respectful of others and will demonstrate a strong desire to help others, support the organization, and build the team. Success of the organization and success of subordinates, peers, and superiors are high priorities for the fire service leader who exhibits socialized power orientation.

Leaders with socialized motivation oppose the manipulation of others. They will use formal power and authority only as a last resort if all noncoercive ways to influence followers have been determined to be ineffective. Nevertheless, a leader with socialized motivation will never shy from taking any and all appropriate disciplinary action when required.

Socially oriented leaders make great efforts to align the needs of the firefighters with the goals of the formal organization. They focus on coaching and counseling to mentor, motivate, and encourage subordinates (fig. 7–1). Fire service leaders with socialized motivation will listen openly to feedback, input, and advice from others. Socialized leadership can be described as servant leadership.

Socially oriented leaders are not pushovers or soft touches. They understand the roles and responsibilities of the people under their command. In addition, they are clear as to their own formal responsibilities and will not hesitate to take actions to insure compliance and to carry out the mission.

Socialized leadership should never be confused with casual, absentee, weak-kneed, or laissez-faire leadership. Socially oriented leaders' success and effectiveness are totally dependent on their ability to assess and react to the dynamics at work among the personnel assigned to their command. Honesty, empathy, fairness, consistency, and emotional maturity are hallmarks of a leader with socialized motivation. What may look to some as weakness or timidity is actually great strength and personal courage.

Fig. 7–1. Training is a form of socialized leadership in action. (Courtesy Al Pratts)

Socialized motivation allows fire service leaders to create environments that stimulate and motivate. A stimulated and motivated fire force will be engaged and mission friendly. One consequence (intended or unintended) of socialized motivation and servant leadership is improved morale. The communication channels in a fire department with socialized leaders tend to be fluid and therefore effective. Subordinates are encouraged to provide feedback, show initiative, and solve problems independently. Thus, employee skills and talents are more readily discovered and more fully developed.

Socialized orientation encourages and values learning, initiative, and inquiry. The loyalty of followers is often bestowed on a leader they respect and the leader's goals and vision for the unit and organization (fig. 7–2).

Socialized power orientation propels forward progress and secures the leadership future of an organization. As a result of focused subordinate development, the organization remains cohesive and continues to function well when a leader leaves the position. A key responsibility of servant-leaders is the creation of future leaders, including their own replacements.

Fig. 7–2. A young Mayor Turner of Weehawken promotes an even younger Anthony Avillo to captain. For me, promotion was the moment where the belief that I could do the job better than my boss crashed up against the reality that I had to do the job that my boss was doing.

Formal and Informal Power

Those with formal power exercise it on a daily basis in the process of doing their work. Those with informal power only exercise it when it suits their personal interest.

—anonymous

Power means many different things to different people. Power comes in different forms, and leaders need to learn how to handle each type. The exercise of different types of power affects leadership ability and determines success in a fire service leadership role. There are seven bases of power that are divided among two categories: formal and informal. *Formal power* flows from the formal organization (the fire department). Formal power is where a person in a higher position has control over employees in lower positions in an organization.

The fire service assigns formal power in accordance with rank, position, and the authority associated with that position. Formal power requires that members of the organization recognize the authority and responsibility assigned to each supervisory level. A chain of command described in a scalar table of organization identifies the levels of power and authority in a fire department.

Informal power flows from relationships built and the respect earned from co-workers. Informal powers are not dependent on rank or organizational positioning. Although the organizational structure of fire departments determines and assigns formal power, personal perception is the engine that drives the assignment of informal powers.

Seven Types of Power

Rank does not confer privilege or give power. It imposes responsibility.
—Peter F. Drucker

Legitimate power (a formal type of power) comes from the position a person holds. This is related to a person's title and job specifications. Legitimate power is also known as "positional power."

Coercive power (a formal type of power) is exercised as threats of dismissal, demotion, reassignment, reprimand, charges, and other punitive actions. Coercive power is unlikely to win respect and loyalty from employees. Leaders heavily dependent on coercive power cannot expect to build credibility with subordinates and co-workers. Leading through coercive influence is essentially leading by bullying.

Reward power (a formal type of power) is conveyed through rewarding individuals for compliance and initiative. Examples of reward power are raises, promotions, preferred assignments, and time off from work. Reward power is effective only if employees believe that the supervisor can deliver the kind of reward promised or expected. Fairness and equitable demonstrations of reward power are necessary to give validity to this form of power.

Expert power (an informal type of power) comes from one's experiences, skills, or knowledge (fig. 7–3). Expert power flows from the perception that a leader possesses superior skills or knowledge. When a leader is considered an expert, others will respect that leader's direction and opinions. Subordinates are more likely to follow the lead of an expert. Expert leadership tends to be recognized on a situational basis. To support a subordinate's perceptions and maintain status and influence, experts need to continue learning, improving, and demonstrating their expertise.

Fig. 7–3. Expert power in action at a Firefighter 1 training session. You can't go to the chief's office and get some expert power. Expert power is assigned by those being led.

Connection power (an informal type of power) is where a person attains influence by gaining favor or simply acquaintance with a powerful person. This power is all about networking. If X has a connection with Y and Z wishes to get next to Y, then that gives X power. People employing this power seek to build and maintain important coalitions with those in power. Connection power is, in reality, power twice removed. Connection power relies not on the person with the connection but with the person who is the connection. In essence, connective power is totally dependent on the good or ill intentions of a third party.

Informational power (an informal type of power) is a type of power where a person possesses information that is needed, wanted, or deemed critical by those being led. This is a short-term power base that does not necessarily heighten influence or generate credibility. Informational power is never a good long-term strategy because it dissipates as soon as the information is disseminated throughout the group or organization.

It is lost whenever the information is determined to be irrelevant. Informational power often arises in informal group settings where the conversations are grapevine, scuttlebutt, and gossip driven.

Referent power (an informal type of power) is the ability to convey a sense of personal acceptance or approval. Referent power is held by leaders with socialized orientation, competency, compassion, intelligence, emotional maturity, charisma, integrity, and other positive qualities. Referent power is the most valuable type of power because it flows from the respect and trust of those being led. Referent power exists when others trust what the leader does and respect how the leader handles situations. Referent power reflects the desires others have to identify favorably with the leader or with what the leader symbolizes to them. Leaders possessing high referent power can have great influence over those who admire and respect them. In addition, they can influence subordinates, peers, superiors, and the organization.

As these types of power show, you do not have to be in a formal leadership or senior officer role in a fire department to have power. In fact, the most respect is garnered by those who have informal sources of power (referent, expert, or both). There is more respect for these individuals than for those who have and use power simply because they are the boss.

It has been shown that when employees in an organization associate a leader with expert power or referent power, they are more engaged and more loyal to the organization, as well as to their role within it. Those being led are more willing to go the extra mile to support and reach organizational goals. No fire department can assign referent power or expert power. Both expert and referent power are hard-won power bases that must be earned on a day-to-day and situation-by-situation basis. It should be made clear that having power is not the same as having rank. Having referent or expert power (or both) does not give license to ignore or disregard the orders or directions of superior officers.

Leadership Styles and Types of Power

Power resides only where men believe it resides.

—George R. R. Martin

Types of power and styles of leadership are themselves tools and skills. Thus, to be effective, leaders must possess an informed, functional, and empirical understanding of leadership styles and types of power. Recognizing when power needs to be exercised and projected, along with knowing which types of power are appropriate to a particular leadership style, allows a leader to successfully navigate the treacherous waters of command.

Effective leaders do not rely on a single leadership style, nor should they depend on or project any single type of power. Consciously or unconsciously, most leaders aspire to acquire, exercise, and project expert and referent power.

Only in the Fire Department of Utopia can a leader even dream of an infinite supply of referent power. Successful leaders will exercise and project all types of power and every style of leadership as necessary.

The dynamic between power and leadership style manifests during every leader-follower transaction. Leadership style is inexorably linked to power. The use of one type of power or leadership style does not prohibit other forms of power and leadership styles to exist in the same leader-follower transaction. It is common for a combination of power types and leadership styles to be present in any leader-follower transaction. Context is the key.

The four styles of leadership depend on particular types and different combinations of power. Certain types of power are compatible with different leadership styles.

The fire and emergency ground allows for fairly straightforward assessment of power and leadership. The consequences of powerless leadership can be severe and even tragic, and they are avoidable.

On a fire or emergency scene, the dominant leadership style will be autocratic. Whenever autocratic leadership is exercised, legitimate, expert, reward, and coercive power are implied or projected either in combination or all together.

Referent power is a personal power that is recognized by and bestowed on a leader by those being led. Often the level of referent power leaders are granted by those under their command can be used to predict how much and for how long fire ground operators will extend themselves for that leader.

On the fire and emergency ground, there is no place for ambiguous, confused, or tentative leadership (fig. 7–4). Autocratic leadership in proper combination with appropriate power bases reduces the number and type of decisions emergency ground operators are required to make. Fire ground commanders who create an environment that allows for too many options will lose control of their forces, get people hurt, and lose more buildings.

Fig. 7–4. Command success is based on specific, well-communicated, and total-immersion leadership. (Courtesy Ron Jeffers)

SOPs are specific prescriptions for action that are designed to focus the fire force on the tasks at hand and reduce any chance for confusion. SOPs essentially represent codification of autocratic leadership, legitimate power, expert power, reward power, and coercive power.

In nonfire and nonemergency environments, threats to life and damage to property are not generally present. Still, it should be understood that all soft-environment activities can—and often do—have a significant impact on harder environments. Therefore, the consequences of weak-kneed leadership and misapplied power in soft environments should not be ignored, condoned, or left without remedy. The potential fallout from weak-kneed leadership and misapplied power in the soft environment includes, but is by no means limited to, breakdown in the chain of command, dilution of command unity, flawed communication, delays, do-overs, bruised egos, administrative kerfuffles, disciplinary action, morale issues, feuds, damage to equipment, poor scheduling, imbalanced division of labor, and poorly maintained apparatus, equipment, facilities, etc., and ad infinitum. None of these consequences is desired or acceptable, and all of them are avoidable. Informed full-contact leadership and properly motivated use of power are the engines that drive the organization to successful mission accomplishment.

The Peter Principle and the Fire Service

With great power often comes great confusion.

—Dan Allen

Most career fire departments promote on the basis of results of standardized testing. The individual at the top of the list is the first person considered for promotion, the second is considered next, and so on. The test administration process, including the testing itself, grading, and monitoring, may be under the auspices of a state civil service department, the individual department, or some other authority.

Such competitive testing is not the only method or metric used to determine promotion eligibility. In volunteer departments, members can be elected to officer and command positions. In some career departments, a "chief's test" is administered. Because our experience is with competitive promotional testing, we will limit the discussion here to that form of promotional advancement.

The Peter principle is the notion that members of a hierarchal organization are promoted to their level of incompetency. The concept entails that an employee is promoted on the basis of their performance in the lower position, not the position to which the employee is promoted. Though the underlying philosophy is harsh, the Peter principle is an accurate assessment of the rationale driving the promotional process.

An argument can be made that the Peter principle may not apply to the fire service. Advancement up the ranks is predicated on a testing process. Promotional exams are based on accepted job specifications and the roles and responsibilities of the higher position being tested for. Again, the candidate with the highest score is the first to be considered for promotion, the second highest is considered next, and so on.

Although it can be asserted that standardized testing is one of the fairest and most objective processes for determining a person's qualifications, competency, and aptitude for a position, the truth is both that it is and that it isn't. Whatever the format (multiple choice, essay, or fill in the blank), standardized testing can measure only knowledge of a particular subject matter. By contrast, performance assessment allows for limited evaluation of job-related abilities. Thus, combining written testing and performance assessment provides a larger (but still limited) metric for assessing applicants more broadly, on their knowledge and abilities.

For most promotional candidates, preparing for a promotional examination is like having a second full-time job while simultaneously studying for an advanced degree. A comprehensive study program exposes firefighters and officers to volumes of information, and no one who has expended time and energy studying the fire sciences and arts is ever worse for it. Quite the opposite, students of the fire business are always better for it. Where are we going with this, you might ask?

There are numerous examples of firefighters that have done exceptionally well on tests but still end up struggling at their new position. Firefighters transitioning from company member to company commander often experience a rude awakening in the face of true culture shock. Firefighter to company officer is a leap from being led to being the leader. The skills, competence, and temperament required at the company officer level are far different from what is needed at the firefighter level. Company officers are promoted from a level that required little or no supervisory experience. It is at this first supervisory level that a comprehensive, intensive, and continual fire officer training and development program must begin.

The company officer position is the most pivotal position in the fire service. The company officer position is the make-or-break point in every fire department. Company officers are the face and voice of the department.

Company officers translate into action all the policies, procedures, and directives issued by the formal organization. The company officer ranks are where all chief officers come from. Weak company leaders do not magically morph into strong chief officers; rather, they merely bring their weakness to the next higher level.

Competency at every link in the chain of command should be the preeminent responsibility of the fire department. Accountability for the competency of all officers must be laid at the door of the chief of department. It is incumbent upon every fire department to dedicate time, money, energy, and support for intensive and continuous company and command officer training. Life-safety benefits, a favorable cost-benefit ratio, operational effectiveness, and a high level of professionalism are the positive returns possible when the fire service embraces and invests in a structured and continuing company and chief officer training and development program (fig. 7–5).

Company and command-level officers literally have the power to make life-and-death decisions. Placing the power to make life-and-death decisions into the hands of any individual demands the fire department to invest money, time, and energy in support of career-long programs dedicated to officer training and development. The office of the chief of department should be commended for an officer corps that is professional and effective, and the chief of department should be held accountable for poor performance at the officer level, as well as for any or all failures by the officers to supervise those in their command.

Fig. 7–5. Full-contact company officers are the bedrock of a great department. (Courtesy Ron Jeffers)

Proactive and Reactive Application of Power

With great power . . . comes great need to take a nap.

—*anonymous*

As with much else in life, being proactive is better than being reactive in the fire service. Informed and properly motivated power applied proactively has a greater and more positive impact on firefighters, fire officers, and the fire department than does the reactive application of power.

The fundamental potency of proactive power flows from unambiguous statement of expectations. Power is diluted whenever there is question as to what is expected of the person, unit, branch, division, or department. Power applied proactively and judiciously is always more efficient, more economical, more productive, and essentially much more muscle than reactively applied power.

Proactive power takes the form of written orders, policies, SOPs, concise verbal direction, training, and structured and informal counseling and coaching. Proactive power looks and sounds like clear and unambiguous direction. Proactive power is concise, comprehensive, functional and practical. All forms of power are finite, and all power has a shelf life and expiration date. Every form of power is constrained by practical, formal, and legal limiting factors.

Reactive power is neither good nor evil; rather, it is a tool. No organization can predict or prepare for every situation, condition, or exigency. Reactive power needs to be exerted when discipline, training, orders, procedures, or policies need reinforcement or redress.

For example, SOPs are designed to deal with 95% of the conditions and situations encountered on the fire ground; the other 5% fall within the domain of reactionary leadership. A fire service leader who understands, is educated in, and respects the power, authority, and responsibilities of company or chief officer has the opportunity to earn the trust and respect of the women and men who do society's most dangerous work in extreme and life-threatening environments.

chapter 8
COMMUNICATION

The single biggest problem in communication is the illusion that it has taken place.
—George Bernard Shaw

When the trust account is high, communication is easy, instant, and effective.
—Stephen R. Covey

Effective communication is verbal speech or use of other methods of relaying information to convey a point. One example of effective communication is talking in clear and simple terms. Another example of effective communication is when the person you are talking to listens actively, absorbs the information, and understands your point (fig. 8–1).

Fig. 8–1. Battalion Chief Steve Quidor and Deputy Chief Avillo discuss postcontrol operations face to face. Communication requires listening and comprehension. If subordinates do not understand what you say, they cannot do what you say. (Courtesy Ron Jeffers)

Communication is the process of exchanging information in the form of messages transmitted between a sender and a receiver. Symbols, signs, body language, facial expressions, and tone of voice are some of the methods by which communication is achieved. An effective communicator respects that communication is an immersive experience. Humans are assaulted with a continuous barrage of information. A message sender is always competing for a receiver's attention against a storm of ambient data in the background (fig. 8–2).

Fig. 8–2. An ambient data storm blows onto the fireground. (Courtesy Ron Jeffers)

Leading men and women in a business as complex and deadly serious as fire, rescue, and emergency response and mitigation requires leaders to be able to communicate at the highest levels of competency. A good communicator needs a good communication strategy, and a strategy is only as good as the tactics employed to achieve the desired goal. A strategy is a careful plan or method for achieving a particular goal, usually over a long period of time.

Just because you talk a lot, speak loudly, bang a gong, or toot your horn is no guarantee that you are being heard. What happens when a leader falls in a forest and there are no followers around—does the leader make a sound? Communication is the fire department's circulatory system.

> *More than anything a leader is a communicator. A good communicator can be a good leader. A great communicator can be a great leader.*
>
> —Battalion Chief Mike Hern

As mentioned already, exchanging information in the form of messages relies on symbols, signs, body language, facial expressions, and tone of voice. Thus, the configuration of a room, as well as the environment where the communication takes place, will affect the way the message is expressed and received. Regardless of the medium used to send a message, zero communication transpires if the receiver does not clearly hear or understand the message being communicated.

Saying something, writing something, or pointing to something does not meet the criteria for effective message sending unless what is said, written, or pointed at can be heard, seen, and comprehended. Communication is a marriage of sender and receiver; you can't have one without the other.

Leadership demands interpersonal interaction. Interaction between persons and the act of sharing information is communication.

Some basic ways in which leaders communicate include speech, signing, hand signals, semaphore, body language, facial expression, eye contact, and touch. Dress codes, office assignments, work assignments, job specifications, appearance, and demeanor are all forms of communication that provide context and often speak louder than words.

Communication is used to transfer information from one entity to another. Communication always involves a sender and a receiver. A communication cycle is complete only when the receiver has understood the sender's message and intent.

Communicating information, messages, opinions, and thoughts can be accomplished with an assortment of communicative aids such as books, the Internet, smartphones, demonstrations, film, portable radio, public address systems, and all forms of media.

Verbal Communication

There are four types of communication: *verbal, nonverbal, written,* and *visual*. Verbal communication includes sounds, words, language, and speech. Speaking is an effective way of communicating and allows us to express our emotions in words.

Verbal communication is further classified into four types:

- *Intrapersonal communication.* These are the conversations we have with ourselves. During intrapersonal communication, the roles of sender and receiver are juggled while processing thoughts and actions. Intrapersonal communication can be conveyed, shared, or held to ourselves as thoughts.
- *Interpersonal communication.* This is one-on-one conversation. Two individuals will exchange sender and receiver roles to communicate in a clear manner.
- *Small-group communication.* This takes place when there are more than two people involved. The number of people involved will be small enough to allow each participant to take part in the discussion. Unless a specific issue is being discussed or there is a particular reason for discussion, small-group discussions can become chaotic and difficult for participants to interpret. With small group communication, reception of information is often hindered because message receivers are often consumed with the construction of their next statement or response. Miscommunication and misunderstanding are not uncommon during small-group communication.
- *Public communication.* This takes place when one individual addresses a large gathering of people. There is usually a single sender of information and several receivers who are being addressed. Lectures and speeches are examples of public communication.

Nonverbal Communication

Nonverbal communication conveys the sender's message without the use of words or sound (fig. 8–3). Nonverbal communication is the most primal form of communication. Nonverbal communications are physical forms of communication, such as tone of voice, touch, and expressions. Hand gestures, symbols, and sign language are also used in nonverbal communication.

Body posture, facial expressions, and body language convey nonverbal messages even during verbal communication. Similarly, folded arms and crossed legs are defensive nonverbal messages. Shaking hands, patting, and touching are ways to express feelings of intimacy, and facial expressions, gestures, and eye contact are all different ways to communicate. Creative and aesthetic nonverbal forms of communication include music, dancing, and sculpting. It is not uncommon for the message of nonverbal communication to run contrary to what is being communicated verbally. Saying yes while shaking the head to indicate no would be a simple example.

Fig. 8–3. Nonverbal communication supersedes all other forms of communication. In this example, a gesture conveys how the team is doing in the exposure. Afterward, verbal communication could be used to address the chin-strap violation. (Courtesy Bill Menzel Photography)

Written Communication

Written communication is the medium through which the message of the sender is conveyed with the help of written words. General orders, SOPs, training schedules, journals, e-mails, reports, and memos are forms of written communication. Written communication is indispensable to informal and formal communication. The sender can edit and formalize written communication. Written communication can be edited before it is communicated to the receiver. Written communication combines aspects of visual communication when messages are conveyed through electronic devices such as laptops, smartphones, and visual presentations that involve the use of text or words.

Visual Communication

Visual communication involves a visual display of information (fig. 8–4). The message is conveyed through the use of visual aids. Badges, logos, patches, photographs, signs, symbols, posters, and banners help the viewer to comprehend the message visually. Movies, plays, television shows, and video clips are all forms of visual communication. Visual communication also involves the transfer of information in the form of text received through an electronic medium such as a computer or smartphone.

Icons and emoticons are a form of visual communication. When icons are used in a public place, on a phone or computer, they instruct the user about meaning and usage. The greatest example of visual communication is the World Wide Web, which communicates with the masses, using a combination of text, design, links, images, and color. All of these visual features require a receiver to view a screen to understand the message being conveyed.

chapter 8 | Communication 79

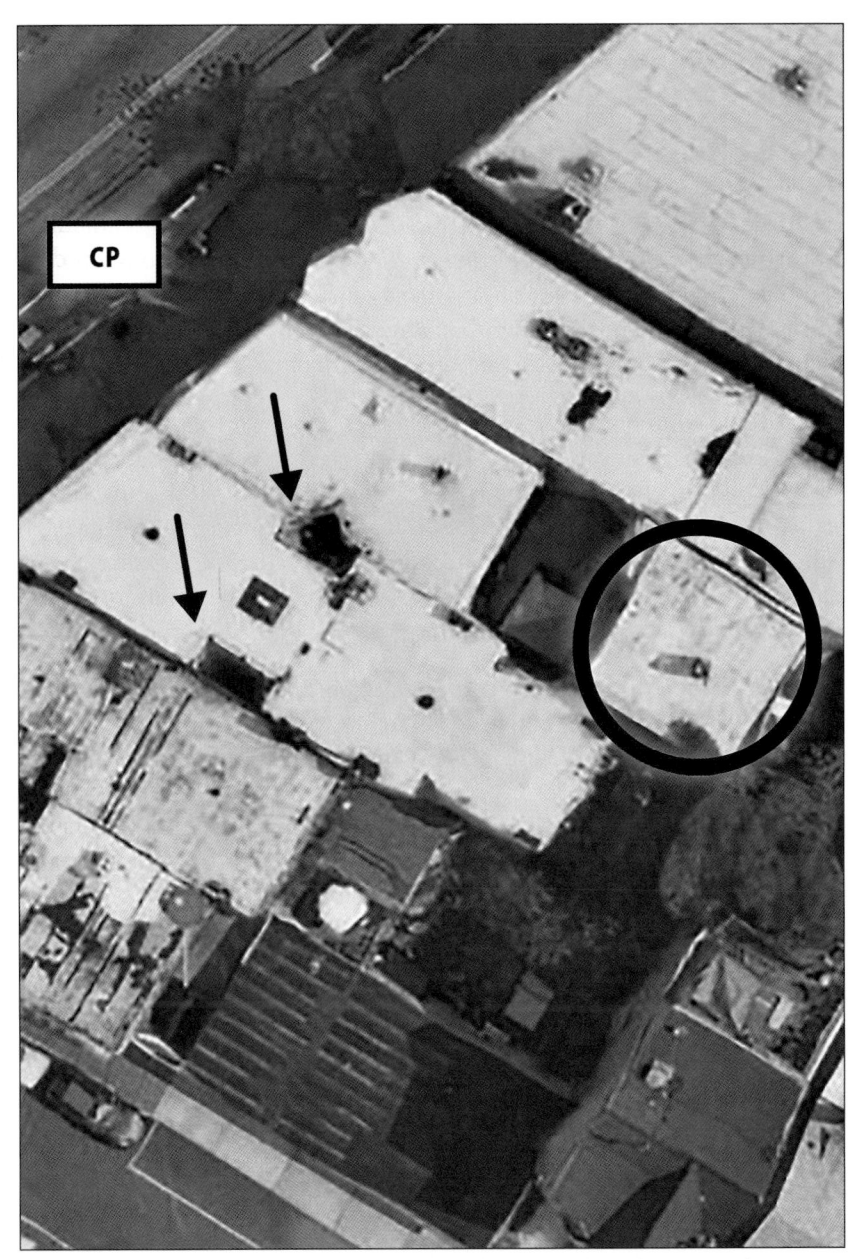

Fig. 8–4. Technology allows fireground information to be viewed on a touchscreen. A cell phone, a map app, and you're in business! Introducing new technologies enhances the effectiveness and extends the reach of command. This image shows a building behind the buildings (*circle*) and light and air shafts (*arrows*). The command post is blind to these features and building configurations.

Hard-Environment Communication

The time, energy, and money that a fire department expends on each individual firefighter and fire officer represents a sizeable investment in its greatest assets. The fire service directs the most energy and attention toward supporting and protecting firefighters and officers during their response to and mitigation of emergency and fire incidents. Fireground communication is a discipline that imposes control and structure onto conditions that are unstructured and uncontrolled.

Hard-environment communication takes the form of SOPs, preplanning, intensive training, tools, equipment, and apparatus—all brought to bear within a highly effective and efficient command and control structure. Fireground communication is effective, efficient, and successful. We know this because, on a routine basis, fires are extinguished, emergencies are mitigated, people are saved, property is conserved, and firefighters go home in the morning.

Fireground communication is successful because it is formalized, standardized, redundant, repetitive, consistent, and contextual. Fireground operators rely on and relay massive amounts of data at emergency scenes. Fireground operators are conditioned to listen to, interpret, and respond to a wide variety of information delivered across multiple communication modalities. Fireground communications include radio transmissions, verbal orders, hand signals, indicator and warning lights, sounds, tones, color codes, and signs. Media displayed on apparel, helmets, and apparatus are forms of communication. Equipment placement and apparatus positioning can communicate valuable information to fireground operators. Firefighters operating at emergency scenes process much of this information reflexively. Fireground operators are conditioned to see, hear, and interpret emergency communication with great speed and accuracy.

The fire, rescue, and emergency business demands that officers and firefighters come together in environments that are by no means communication friendly. Firefighters and fire officers are required to perform complex and dangerous tasks under the most hazardous and stressful conditions. Very few people have the capacity to extemporaneously communicate complex directions under loud, chaotic, and life-threatening conditions.

Much fireground communication takes place long before anyone gets to the fire or emergency scene. Good fireground communication is a function of solid SOPs, preplanning, policies, training, and equipment. If responders were required to improvise on arrival at the scene, every fireground would be a place where firefighters gathered to become an audience for the "Self-Extinguishment by Means of the Thermal Elimination of Combustibles" show.

SOPs, preplanning, policies, training, and equipment comprise 95% of all fireground communication. Fireground communications announce the strategy and jump-start tactical action. Communication on the fireground has to do with confirming that the actions taken and procedures followed are achieving the desired results. Fireground communications provide for personnel and material support, enhanced supervision, reporting of exceptional conditions, and facilitating de-escalation and cessation of operations.

The fire service arrives at every incident with a portfolio of prepared and standardized statements, announcements, and protocols. The communication portfolio is designed to contend with just about every possible scenario that could be imagined or predicted. Emergency response communication has an inherent rigid flexibility that allows effective operational communication to adapt to new, unknown, and unforeseen conditions. Effective fireground communication is standardized, clear, concise, and readily understood.

Words at Work

Formal communicating takes place regularly between leaders and those being led. However, informal and semiformal conditions are the most common venues for interaction between leaders and the firefighters assigned. There is no leadership if there is no communication. Leadership cannot be successful without effective communication (fig. 8–5). Even with that knowledge, not much forethought appears to be given to any but the most formal communication scenarios. Little conscious preparation is directed toward routine, day-to-day soft-environment communication.

Fig. 8–5. Communication skills are as important as or more important than any other tool available to leaders in the fire service. (Courtesy Len Carmichael)

Ninety-five percent of on-duty time is spent under soft or semisoft conditions. The remaining five percent is dedicated to fires, emergencies, rescues, and a variety of other situations that require PPE.

There is a school of thought that is okay with extemporaneous communication. This school holds that the only place good communication truly matters is on the fire or emergency ground. Many proponents of the "it only matters on the fireground" attitude believe great emergency ground communication skills carry over to soft-duty communication. The converse is true, however: great communicators in soft environments are often the best emergency ground communicators. Soft environments offer a leader the opportunity to hone communication skills. "You play like you practice" is a truism that resonates throughout the fire service.

Other than perspiration or breath, communication is the most abundant by-product of human endeavor. For the most part, humans do not need to practice breathing, and for some reason, they believe that it is not necessary to practice interpersonal communication except on the most formal occasions.

For effective execution, interpersonal communication should be studied, practiced, and prepared. Clear, concise, and complete—the *3C model*—are the three requirements for effective fireground communication.

Communications taking place off the emergency ground should also aspire to the 3C model. Every word a leader says has impact. Words should not be sprayed around like insect repellent.

Fire service leaders often confront issues in soft environments with relatively smaller communication skill set and with less preplanning. When compared to fireground communication, soft-environment communication seems long winded and ambiguous. Such communication usually requires the receiver to decode the message being relayed. Other times soft-environment communication takes the form of sighs, head shakes, smirks, and knowing smiles—and may even be combined with guttural noises that harken back to our primitive Neanderthal roots (I may have exaggerated to make a point here).

A fire service leader is under constant scrutiny. Every word and every action or inaction is reviewed, interpreted, and filed in the collective consciousness of the group. Fire service leaders must understand the impact of and respect the power of their words. Words must be well considered and apportioned with discretion and purpose.

In the hard environment, errors of judgment, failure to comply, disregard for procedure, and just about any mistake, great or small, can immediately jeopardize life and result in an infinite number of tragic consequences. The dangers associated with emergency operations are readily observable to members of the fire service and should receive appropriate attention and support. However, support for the response and mitigation of soft-environment communication issues is often considered less important, and consequently the support given is not proportional to the department's investment or the amount of time firefighters and officers spend not involved in hard-environment action.

Consequences of communication errors confined to the soft environment do not usually present immediate threats to life or property and are rarely tragic or extreme. Nevertheless, lapses in discipline with regard to soft-environment communications can lead to bigger problems, resulting in issues that are harder to resolve and actions that involve higher ranks. Communication lapses in soft and semisoft environments routinely come home to roost on the fireground.

Often the resource drain from nonemergency time and activities goes unrecognized or is considered less worthy of support and attention. The intensity and danger associated with firefighting overshadows and diminishes concern for what takes place during soft time.

Fire service leaders lacking in communication skills are often reluctant to intervene in seemingly minor issues or infractions in soft environments. However, such initial reluctance forces the officer and the organization to play catch-up as issues continue or escalate. Positive discipline, in the form of effective communication, usually serves when intervention is immediate.

Fireground versus Soft-Environment Communication

Soft-environment communication involves written and verbal orders, formal and casual discourse, phones, intercoms, public address systems, tones, announcements, bulletin boards, apparel media, color coding, unit identification, and markings for tools and equipment (fig. 8–6). Even the floor plan, the type and positioning of furniture, and office and room assignments are forms of communication.

Fig. 8–6. Forms of communication are everywhere—from the patch, to the meritorious service medals, to the Class A uniform. Even the salute is a means of communication. (Courtesy Ron Jeffers)

Much of the fireground communication model can be modified and effectively applied to soft and semisoft communications. The soft environment is where the formal and informal groups mingle, blurring the hierarchical boundaries of the organization. For good or ill, soft-environment behavior always has the potential to require the time and attention of the formal organization. An effective leader will recognize that most soft-environment activities are routine, recurring, and predictable.

An effective leader should develop and preplan responses for reference and potential use during routine or anticipated situations. A leader armed with standardized soft-environment responses does not have to improvise every conversation and has tools to preempt unnecessary and inappropriate incursions across the boundaries of the formal organization.

Transactional Analysis

The art of communication is the language of leadership.

—James Humes

The difference between the right word and the almost right word is the difference between lightning and the lightning bug.

—Mark Twain

Transactional analysis is a method to determine the involvement of ego in human behavior and communication between individuals. Transactional analysis is based on two notions: first, that we have three *ego states* comprising parts of our personality; and second, that these ego states converse with one another in *transactions*. Transactional analysis identifies three ego states: parent, child, and adult. The behavioral influence of these ego states varies in strength at all times. In transactional analysis, each communication between people is considered a transaction, and these transactions are driven by the dominant ego state. Transactions are identified as *complementary, crossed,* or *hidden.*

Parent ego state

Acts like: angry or impatient body language and expressions, finger pointing, patronizing gestures

Sounds like: always, never, once and for all, judgmental words, critical words, patronizing language, posturing language

Child ego state

Acts like: emotionally sad expressions, despair, temper tantrums, whining, rolling eyes, shrugging shoulders, teasing, delight, laughter, speaking behind hand, raising hand to speak, squirming, and giggling

Sounds like: baby talk, I wish, I don't know, I want, I'm gonna, I don't care, oh no, not again, things never go right for me, worst day of my life, bigger, biggest, best, many superlatives, words to impress

Adult ego state

Acts like: attentive, interested, straightforward, tilted head, body language that is nonthreatening and nonthreatened

Sounds like: why, what, how, who, where, when, how much, in what way, comparative expressions, reasoned statements, true, false, probably, possibly, I think, I realize, I see, I believe, in my opinion

Using transactional analysis

To analyze a transaction, you need to see and feel what is really being said:

- Only 7% of meaning is in the words spoken
- 38% of meaning is *paralinguistic* (the way that the words are said)
- 55% is in facial expression and body language

Transactional analysis is the language within the language and can be used to find the true meaning, feelings, and the motives that are driving the communication. Understanding transactional analysis enables leaders to make the most of face-to-face communication. Transactional analysis can go a long way toward creating healthy and productive work environments by allowing a leader to clearly see the situation for what it really is. The leader is then given choices of which ego state to adopt, which signals to send, and where to send them.

Case Study: Transactional Analysis the Hard Way—As Avillo Saw It

Communication is the real work of leadership.
—Nitin Nohria

Skill in the art of communication is crucial to a leader's success. He can accomplish nothing unless he can communicate effectively.
—anonymous

I was a ladder officer in a double-company house. The crew had recently replaced a rack where the pots and pans were kept. The new rack consolidated the huge inventory of cookware we had in the firehouse kitchen. It wasn't much of a change, but it made the kitchen a little more livable. As we were basking in the glow of our handiwork, our platoon commander, Battalion Chief Ed Flood happened to walk in. The engine officer brought the commander's attention to our brand-new pot-and-pan rack. Battalion Chief Flood acknowledged our good work, saying that it looked much cleaner and more organized.

As he often did, the chief urged us to go a step or two further. He pointed out a number of other steps that could make the kitchen even more livable. Most of the tasks he directed our attention to would require less effort than the great pots-and-pans project. Still, a couple of the things he pointed out felt like a heavy lift and as I watched and listened to the chief point these things out, I began to work up a considerable amount of righteous indignation. How could the chief disregard the herculean effort we had extended?! Of course I could not stop my mouth from doing my thinking for me and blurted out my objections with some choice profanity for emphasis and persuasion!

A quiet, cold dark wind seemed to blow through the room, and time seemed to stand still. Somewhere in the distance a dog barked, and everyone present in the kitchen did a synchronized jaw drop. I sensed my outburst may have placed me smack in the middle of a free-fire zone—or a Flood zone!

I began to think that I had made a dreadful mistake. I could not have been more correct. Battalion Chief Flood did not immediately look up from whatever it was he was looking at. When he did, he laid a hard, dead-eyed look on me and politely repeated verbatim my question, curse words and all. Then he asked if that was in fact the question I had posed.

I began stammering a torrent of feeble excuses. Battalion Chief Flood then repeated my original question twice more. Finally I reigned in my flapping jaw and shut my mouth. The chief had basically questioned me into silence. I was certain the other firefighters in the room were not going to remember this as my greatest command performance.

After I had ended my not very effective defense, the chief graciously invited me to meet him his office in five minutes. He said he would be happy to answer my original question and discuss any other issues I might feel the need to explore.

In his office, the chief told me that he had just done me a very great favor. I wondered where the conversation was going to go. "Captain Avillo," he began, "you were shooting yourself in the foot with that ill-advised rant." He explained that the action I modeled in the kitchen undermined my own stature as a leader. Battalion Chief Flood explained that it would have been unbecoming for two officers to engage at the place where my emotional outburst was coming from. Such behavior would speak poorly to the firefighters and the officer who stood witness to the interaction. He told me it was his responsibility to save me from myself. He did what he did not only for my sake but also for the sake of the firefighters under his and my command.

"The people who work for you will adopt the attitudes and behavior you exhibit." Essentially, he was explaining that if I was allowed to run my mouth off at a chief officer, then I would be giving permission for the people who worked with and for me to run their mouths off at me and other officers. If the chief had joined me in the exchange, then we both would have set a very poor example. An argument rarely solves anything. Mature and honest communication is how problems are resolved and issues confronted (adult-to-adult transaction).

Before he released me and my substantially bent ear, the chief gave me an assignment. He told me that I was to prepare and present a company-level training exercise. The topic I was to teach was transactional analysis. I did and I learned, and now I will forever know it when I see it.

The reader is invited to analyze and discuss how and why transactional analysis applies to the scenario described above.

chapter 9
DISCIPLINE

Discipline isn't a dirty word. Far from it. Discipline is the one thing that separates us from chaos and anarchy. Discipline implies timing. It's the precursor to good behavior, and it never comes from bad behavior. People who associate discipline with punishment are wrong: with discipline, punishment is unnecessary.
—Buck Brannaman

Discipline is what it takes to be a disciple of success.
—Constance Chuks Friday

If you manage yourself, you control the flow of your time to the right direction. It takes self-discipline to be at the center of control for your own time.
—Israelmore Ayivor

Each one teach one . . . learn something new every day . . . teach something new every day.
—Flood

Discipline Is Teaching and Learning

Discipline creates, supports, and maintains the order and structure of a fire department. When properly administered, discipline levels the playing field by ensuring the fair and equitable treatment of all members.

Discipline should be understood to be a tool designed to bring about positive results. Acts of discipline should be administered as a teaching and learning experience. Disciplinary actions are designed to protect the organization and provide employees with opportunities to contribute to the success of the organization at higher and more rewarding levels.

Discipline Gets a Bad Rap

We have delivered many officer development programs on leadership, management, and supervision. Just the word "discipline" seems to trigger a visceral fearful and hostile reaction. Anytime we have broached the subject, a palpable wave of "Oh no, not discipline—anything but that!" has washed through the classroom.

It is a common misperception that no one, especially firefighters, enjoys being told what to do, how to act, where to go, what time to be there, or much of anything else. The real deal is that firefighters are prouder, happier, perform better, work harder, and find the "JOB" more rewarding when they are well trained and when the expectations, rules, policies, and procedures of the department are clearly defined and respectfully expressed in a loud, clear organizational voice.

The organization identifies where members stand in relation to each other, as well as where they fit into the department. This structure reduces stress and anxiety and creates a stable work environment.

Disciplines are the sinew of an organization. The organization has no muscle without discipline. Discipline is a personal and organizational tool. No single form of discipline is more appropriate than another. All forms and combinations of discipline must be acknowledged and invoked as each situation requires. Fire service leaders must embrace the idea that discipline is a positive word, a positive concept, and an invaluable tool that has the greatest potential for moving organizations, groups, and individuals in a productive, positive, and self-esteem-promoting direction.

Three Types of Discipline

If there was an election to determine everyone's favorite type of discipline, most would vote for *self-discipline* over the two other contenders, namely *positive discipline* and *negative discipline*. This is mainly because self-discipline sounds like it that would allow plenty of wiggle room and little accountability. Positive discipline would come in a close second. This could be because positive discipline sounds so positive and again because there might be a good bit of wiggle room built in. Negative discipline would have slim chance of being be elected as anyone's favorite type. Not surprisingly, this may be because negative discipline sounds so negative and because negative discipline does not sound like a wiggle-friendly form of discipline.

Self-discipline

Self-discipline is the discipline that an individual brings to the table. Self-discipline comprises the self-affirming instincts, the positive and mature behaviors that individuals exercise while going about their life experience. Self-discipline is the foundation upon which other forms of discipline find a foothold. Self-discipline must be supported, reinforced, and empowered with positive discipline. Self-discipline also plays a role in driving adjustment, redirection, and education of oneself after negative discipline.

Positive discipline

Positive discipline comprises the organizational expectations clearly set out and codified in rules, regulations, policies, and training. Positive discipline builds on and compliments self-discipline. Although self-discipline is beautiful thing on its own, when self-discipline is combined with positive discipline, extraordinary results are achievable.

The hard truth in the fire and rescue business is that positive discipline demands rigorous efforts and focused attention. Safe, successful, and effective fire, rescue, and emergency operations demand extremely high levels of positive discipline.

Positive discipline is a productive, cost-effective, and organizationally beneficial mechanism available to fire departments seeking to provide high levels of professional fire, rescue, and emergency service to the community. Positive disciplines are procedures, policies, activities, exercises, regimens, training, and skills designed to enhance skill levels and understanding (fig. 9–1). The goal of fire department learning and discipline is to expand contributive potential, facilitate, and provide opportunities for employees to satisfy self-realization needs.

Positive discipline is a program of affirmative reinforcement, designed to enable members to comply with organizational protocols and standards. Positive discipline fosters appropriate behavior, invites employee investment, and encourages employee participation.

Fig. 9–1. Training is one form of positive discipline. (Courtesy Ron Jeffers)

Positive discipline can be introduced only when clear protocols, ethical guidelines, and codes of conduct are made known to every firefighter and officer. It is the responsibility of the formal organization and the representatives of the formal organization to ensure all members fully understand all department requirements, goals, and expectations. A one-time lecture on department protocols and goals is glaringly insufficient to support positive disciplinary programs. Moreover, believing that any one-time event will achieve disciplinary goals is an overly optimistic and potentially ruinous state of mind. It is necessary to reinforce and review on a regular and as-needed basis any and all rules, regulations, policies, procedures, codes of conduct, and protocols.

Positive disciplines are designed to reduce the need for negative discipline. The exponential return on time, energy, and money invested in positive discipline quantifiably and qualitatively resonates throughout the entire department.

Negative discipline

Negative disciplines are clearly stated, codified, and published actions that prescribe how an employee's behavior is to be realigned. It is used to change employee behavior to conform to the requirements of the service. Negative discipline is used when self-discipline and positive discipline fail to direct, change, or align an employee's behavior.

Negative disciplinary action is punitive in nature. When positive discipline fails, negative discipline is the next step to protect the offender and other members from the unwanted consequences of unsafe or inappropriate behavior. Negative discipline should be designed to have a positive outcome. Maintaining order and preserving the integrity of the fire department is a positive outcome.

Negative discipline may involve any degree of punitive response, from a warning to severe actions such as separation from service or criminal prosecution. Thus, a hierarchy of punitive action can be instituted, and negative discipline actions between a mere warning and termination include oral or written reprimand, voluntary surrender of accumulated time, surrender of regular days off or vacation time (or both), administrative disposition requiring mandatory relinquishment of time off, demotion, loss of salary, or suspension.

Negative discipline is more expensive and less cost-effective than positive discipline. It is a drain on an organization's energy, time, and money. However, in situations where negative disciplinary action is necessary, it must be administered. Otherwise, the formal organization and the informal groups monitor, record, and remember every violation that is allowed to go unchecked.

Punitive action is designed to redirect behavior. With progressive negative discipline, the employee is afforded a number of opportunities to realign his or her behavior with the requirements of the department. When the organization has exhausted all positive and punitive remedies and the negative behavior has not changed, the formal organization has no option other than termination. Aside from criminal prosecution, termination is the most extreme form of negative discipline, and it is the least cost effective.

Whether career or volunteer, a fire department represents a massive investment by a community. The community has every right to expect the fire department to protect its investment. Substantial salary, compensation, and benefit packages are available to career firefighters. Time, attention, and person-hours are dedicated to developing, supporting, and housing a fire and rescue force so that it can provide the highest level of service to the population. Discipline is designed to protect not only the employee but also the substantial investment made by the fire department and the community served. Termination is the most extreme disciplinary option and should be imposed only after all positive and negative disciplinary options have been exhausted.

For any fire department, the individual firefighter is its greatest asset and represents its greatest investment. All negative disciplinary actions should seek to balance what is best for the individual and what is best for the organization. Ideally, when negative discipline is imposed, it will align the behavior of the individual with the requirements of the formal organization. Both negative and positive discipline are safety nets—that is, corrective devices that protect the formal and informal organization, as well as the community served.

The department must never hesitate to use negative discipline in the event of illegal, criminal, or unsafe employee behavior, such as sexual harassment, violence, or failure to follow vital safety protocols. A fire department is required to report all illegal and criminal behavior to the proper authorities. Anytime a fire department is confronted with issues or behaviors that are criminal or illegal, the department should immediately seek legal counsel.

The administration of negative discipline must follow and adhere to all legal and contractual mandates. Haphazard and casual attitudes toward the administration of disciplinary action do a disservice to the department and all parties involved. Faulty disciplinary proceedings create an environment of ambiguity and confusion. Failure to recognize and meticulously follow proper disciplinary protocols will create conditions that have the potential to nullify the action, incur unnecessary costs such as legal fees, and cause further administrative disruption.

All rules, regulations, policies, and procedures are formulated to create safe, productive, and rewarding work environments. An organization is responsible for educating the rank and file about the meaning and true purposes of positive and negative discipline. The way in which a fire department handles negative disciplinary action sends a message that reverberates through all formal and informal groups within the department. No one in the organization, from the chief of department down to the rawest recruit, should find or express satisfaction (gloating) over the disposition of negative disciplinary measures.

Firefighters, fire officers, and fire chiefs are never really off duty. Realistic or not, the behavioral expectations of those who are served by the fire service are high. The effects and consequences of discipline are not confined to the on-duty, physical, or procedural limits of the fire department. Positive discipline represents both the department and fire service as a whole in a positive light. By contrast, negative, illegal, or criminal behaviors engaged in or exhibited while a member is off duty always reflect back on the department and often require extradepartmental remediation or department-level disciplinary action.

Discipline Is a Product of Communication

Learning is a process of active engagement with experience. It can never happen without good communication. Learning is necessary for firefighters to function effectively and make sense of the fire service world. Good communication is the cornerstone of a firefighter's education.

The fire service is one long and continuous learning experience (fig. 9–2). A full-contact leader is both student and teacher. The effectiveness of a fire service leader is almost wholly dependent upon communication skills. Effective learning and discipline should lead to changes in behavior, to personal development, and to the desire to learn more and achieve at higher levels. All of these are accomplished through good communication.

Fig. 9–2. A career in the fire service is one continuous learning experience. Full-contact leadership promotes the concept of a career of learning.

Fire service leaders are responsible for the development of both themselves and their subordinates. The learning experience should involve deepening of skills, knowledge, understanding, and values resulting in positive behavioral change. Fire service leaders must keep in mind that they are only holding a place that belongs to a next generation of leaders. A full-contact leader should actively work to create the full-contact leaders who will take their place.

Discipline is a tool for both improving performance and changing behavior. Discipline protects both the individual and the formal organization, including protecting individuals from themselves. Importantly, all discipline, whether negative or positive, is a learning experience.

You Know It When You See It

If a woman in a plain blue sweatshirt gets arrested, it is not guaranteed to be news. However, if a woman in a sweatshirt with fire department identification and department patches on it gets arrested, that is going to be news. Similarly if a guy in a white t-shirt is filmed acting inappropriately, it may not go viral on YouTube, but if a guy in a fire department t-shirt is filmed acting inappropriately, that will most definitely take off from YouTube and across social media.

You Know It When You See It

chapter 10

BLOCKS TO EFFECTIVE LEADERSHIP

Searching for the best solution is by far the best excuse for inaction.
—Meenakshi Sundaram

Not doing anything can be worse than doing the wrong thing.
—Alexandra Potter

Leadership Thinking

Full-contact leadership requires keeping up one's guard. In the fire service, a casual approach to leadership can kill. In every chain of negative consequences, there are points for corrective intervention. Corrective intervention can bring a halt to dangerous operations and stop unsafe behavior. Corrective intervention has the power to save smart people from the consequences of doing dumb things.

Somewhere in every chain of negligence, a bogus assurance will be agreed upon: "It's okay. Go ahead, it's no big deal"; or "Don't make a federal case out of it"; or "Nobody's going to notice"; or "Take it easy. It's only [blah, blah, blah]."

If you ever find yourself pursuing this line of reasoning, which is in fact ignorance, stop. If you are in the company of any who espouse minimalist emergency ground leadership philosophies, stop them or move, posthaste, to safer ground.

Cognition and Cognitive Premeditation

Cognition. Conscious mental activity; the activities of thinking, understanding, learning, and remembering.

Cognitive premeditation. The cognitive process of thinking what you will do in the event of something happening; planning or plotting in advance of acting; forethought; planning; preparation; provision.

It is difficult for firefighters to stand down, stand by, or just plain wait. The fire service reveres action—doing, accomplishing, responding, mitigating, and rescuing. Firefighters want to take care of business, get the cat out of the tree, get in on the action, make the great stop, and pull the victim out. This attitude of straining at the bit reflects the courage and commitment of the amazing and dedicated people who are the face and backbone of the fire, rescue, and emergency services. The desire to conserve property and save lives despite great personal peril makes firefighting one of the most purely altruistic vocations in society.

Without clear thinking (cognition) and preparation (cognitive premeditation) action is counterproductive and dangerous—often leading to tragic consequences. Leadership is cognition in action, and full-contact leaders do much more thinking (cognitive premeditation) than acting. That is exactly the way it should work. The action-oriented environment of the fire and rescue business creates command and leadership challenges for the officers charged with safely directing, commanding, and controlling the actions of fire ground operators during emergency activities (fig. 10–1).

Fig. 10–1. A full-contact leader uses every opportunity to mentor, teach, guide, explain, and support. Full-Contact leadership is a "get 'em in safe, work 'em safe, and get 'em out safe" kind of leadership. (Courtesy Ron Jeffers)

Unintended Consequences

Unintended consequences are outcomes that are not the ones desired or expected. They can be the result of action or inaction.

Unintended consequences are either positive or negative. A *positive consequence* is an unexpected benefit derived from the action or inaction. Luck is an example of positive unintended consequences. A *negative consequence* is when an intended solution makes a problem worse. Murphy's law—that whatever can go wrong, will—exemplifies this. Possible causes of unintended consequences include complexity, perverse motivation, stupidity, self-deception, failure to account for human nature, failure to account for acts of nature, lack of competency, lack of preparation, and communication breakdowns.

Failure of Imagination

This describes when something seemingly predictable and undesirable was not planned for. Thus, failure of imagination is a situation where predictable and undesirable consequences are the result of purposeful action or inaction. Causes for failure of imagination include a self-confidence vacuum, inability to organize, lack of attention to detail, shortsightedness, and apathy.

Functional Fixity

Functional fixity is an inability to realize that something known to have a particular use may also be used to perform other functions. This blocks one's ability to use familiar tools in novel ways when faced with new problems. Functional fixity can result from routine because doing the same thing over and over can stifle creativity and innovation. People are often very limited in the ways they think about objects, concepts, and people. Functional fixity also presents as a type of thinking that is narrow and limited, often inhibiting the problem-solving process.

Deviation Amplification

Deviation amplification is a process or condition where the initial trajectory of an action is slightly off and the result of the action moves progressively farther from the intended goal. Consequently, the extent and seriousness of the deviation is exaggerated.

You Know It When You See It: Failure of Leadership, Failure of Imagination, Functional Fixity, Deviation Amplification, and Unintended Consequences

Scenario

An aerial ladder pipe was being directed through a fourth-floor window of an ordinary-construction tenement building. The turntable-to-ladder-tip intercom was not working.

The firefighters and company officers did not have portable radios. The chief in charge wanted the stream redirected. The chief ordered the ladder captain to go up the aerial and have the firefighter redirect the stream to get more reach onto the fire floor. The firefighter operating the turntable was on detail from an engine company and had little experience operating an aerial device.

When the captain got into position just below the firefighter directing the stream he hand-signaled for the ladder to be lowered. No standard hand-signal system was in place. The captain had not taken the time to brief the turntable operator as to what he was doing and what he wanted to achieve with the aerial ladder.

The inexperienced turntable operator pulled the control lever, retracting the ladder and trapping the officer's ankle and foot. The captain's ankle was crushed between rungs (fig. 10–2).

Fig. 10–2. The firefighter pictured was victimized by a multitude of failures of leadership. It must be made clear that the greatest of fireground commanders using the best strategy and perfectly executed tactics will be thwarted and become a victim of the insidious rot of organizational decay. Seemingly inconsequential mis-, mal-, or non-feasence(s) committed by the organization turns out to be consequential. Every firefighter, officer, and chief at this fire must be considered a victim. (Courtesy Ron Jeffers)

After-action analysis

- The ladder should never have been placed in service. The ladder was not ready for fire operations.
- There was a failure to secure an alternative communication link between turntable and ladder tip.
- The ladder pipe could have been directed with halyards.
- An inexperienced firefighter was assigned to perform an evolution that required expert-level operator skills.
- The captain did not brief the turntable operator on the intended goals of the operation.
- The operation required coordination and enhanced supervision but had neither.
- Extending or retracting an aerial ladder should not be attempted while operators are on the ladder.
- Placing the life and safety of personnel in jeopardy unnecessarily during exterior defensive operations is reckless, unacceptable, and indefensible.
- Unintended consequences, rampant failure of imagination, functional fixity, and deviation amplification created conditions ripe for calamity and catastrophe.
- The department's inability to foresee this casualty (failure of imagination), as evidenced by the failure to fix the intercom, set the stage for this casualty to occur.

Addressing any one of these factors could have prevented this tragedy.

chapter 11
CASUAL AND SENSUAL LEADERSHIP

I probably told you, but you probably forgot.
—Battalion Chief Frank Nagurka

Sensual. *Relating to or affecting senses other than the intellect.*

Casual. *Without definite or serious intention; careless or offhand; seeming or tending to be indifferent to what is happening; apathetic; unconcerned.*

Full-Contact leadership is a process that influences people and enlists their aid and support to accomplish an assigned task or work toward a common goal. Positive, mission-engaged full-contact leadership is the product of premeditated cognitive thinking translated into collaborative action. Such leadership provides a framework for the proficient, thoughtful, job-savvy, and intellectually courageous men and women who are in command or aspire to leadership positions in today's fire, rescue, and emergency services (fig. 11–1).

The fire, rescue, and emergency services in the United States have the good fortune to be led by experienced, dedicated, and highly proficient company- and chief-level officers. Department-wide support for leadership development embracing a creative and effective leadership philosophy will enhance the command experience for highly motivated, mission-focused fire officers. Full-contact leadership encourages command officers to conscientious, correct, and courageous application of the arts and sciences of human relations, management, supervision, and leadership.

Fig. 11–1. Passing it on and paying it forward: The Tactical Perspectives DVD series team. **From left:** Jim Duffy, Chris Pepler, Anthony Avillo, P. J. Norwood, and Frank Ricci.

In full-contact leadership terms, the polar opposite of premeditated cognitive thinking is casual and sensual leadership. Although they can be individuated, casual and sensual leadership styles almost always work in tandem. Casual leadership is apathetic, whereas sensual leadership is feelings based.

Casual and sensual leadership is devoid of planning. Actions, tasks, and assignments are performed haphazardly. Casual leadership is not mission driven and has little concern for either the task assigned or the operators carrying out the assignment. Sensually motivated leaders will approach their job based on how they feel. Sensual leadership is without definite aims or serious intent. By contrast, casual leaders will exhibit a careless, offhand manner and are more often concerned with people pleasing than the quality of their product.

Casual leadership is apathy on steroids. As the antithesis of full-contact leadership, casual and sensual leadership is grounded in a passive mind-set and an unjustified belief that things will probably be okay and people will find a way to work it out.

Casual leadership affects the organization negatively in a top-down and bottom-up manner. Casual leadership breeds confusion, resentment, and unnecessary stress at all organizational levels—among members of the company, battalion, division, and department.

Casual leadership creates voids in the team dynamic, and nature will invariably fill a void. The problem lies in whether the leadership space will be filled before or after a point of crisis has been reached. A void created by a casual leader will need to be filled by someone or something other than the formal leader. This could be a conscientious member of the team, someone who cares and recognizes that a job needs to be done or a team member needs assistance; alternatively, it might be an informal leader whose agenda is less than mission focused. Until these voids are filled, screwups are to be expected, with the only variable being how big the screwup is and how many people are going to suffer or be inconvenienced.

Casual and sensual leaders speak and sound like:

- "It's not my fault"
- "I thought [another colleague or anyone else] was supposed to"
- "Oh, you mean now?"
- "I didn't understand."
- "I'll get on it right away, Chief."
- "Are you sure you told me to do that?"
- "Nobody told me."
- "I can't be bothered right now."
- "Catch me later. We'll go over it then."
- And, of course, the classic, "The dog ate my homework."

Casual leaders rarely do any leading unless they get caught not leading. When forced to lead, casual leaders become anxious and stressed out. They are always playing catch-up. Casual orders are vague and nebulous, have little definition, and seem to lack purpose. Most orders given by a casual and sensual leader are seen by subordinates as coming out of nowhere and often lead to unpleasant surprises. A casual and sensual leader is rarely an effective communicator. When a communicator is under stress, orders are issued with little or no forethought.

Casual orders lack definition, having no clear beginning, middle, or end. There is no thought to follow-up. Orders issued casually create ambiguity and rarely accomplish the results intended.

Orders inherently have an exponential effect. A casual approach to issuing or following orders fails to consider the exponential power of orders. Just telling someone else what to do is a careless and detached act. The results of orders casually issued can range from a small detour to a major catastrophe.

It is no secret that poor, casual, bad, and sensual leadership have a negative impact on individuals, companies, battalions, divisions, and the department as a whole. Casual and sensual leadership is a symptom of an organizational malaise. Thus, casual and sensual leadership must be considered a systemic failure. All forms of command and administrative apathy can be traced back to the door to the office of the chief of department. Poor leadership can exist only when the chief of department fails to hold junior and chief officers accountable for supervisory failures. Failure to supervise should never be ignored.

The chief of department must hold junior and chief officers accountable for any failure to supervise. Routine disregard for supervisory failures implies that the chief and the formal organization are okay with ineffective, disinterested, unengaged, and halfhearted leadership.

The chief of department must set the bar of expectations for performance and accountability for every supervisory position within the department. The chief of department must hold every officer to the performance and accountability bar of expectations. This allows the chief to redirect any and all misaligned expectations. Every junior officer, chief officer, and any member acting in a leadership role must be informed, educated, and be known to clearly understand all the roles, responsibilities, and expectations of the leadership position held (fig. 11–2). The chief of department must monitor every level of leadership to ensure that the needs of department members are being served, the mission of the department is being carried out, and the requirements of service to the community are being met.

It is well recognized that fire department chiefs have a lot on their plate. No one is advocating that any chief run around with a magnifying glass looking for clues, sniffing out foul play, or losing sleep over the "Night of the Living Unled." The chief of department is the supervising overseer. The chief is the hypervigilant boss of bosses and therefore responsible for ensuring an effective, productive, and functional leadership system that is in full and constant contact with its members.

chapter 11 | Casual and Sensual Leadership   113

Fig. 11–2. A full-contact chief of department is the ultimate servant-leader. This means being 24/7 committed, accessible, engaged, energetic, enthusiastic, and a "buck stops here" kind of person. (Courtesy Ron Jeffers)

The soft environment provides for a relatively safe and stress-free leadership laboratory. The soft environment offers opportunities to develop enhance and practice premeditated cognitive thinking and communication skills. Leaders must lead not only in the hard environment but equally in the soft and semisoft environments.

An alarm of fire cannot magically change a casual leader into a full-contact leader. Leaders will play the way they practice. Those who do only what feels good and who shoot from the hip or make it up as they go along when operating in soft environments will discover that a casual mind-set reinforces major blocks to effective leadership when operating in hard environments.

Much firefighting activity is counterintuitive. For example, running in while others run out; destroying property in order to save property; directing rescue operations away from the person at the window to the people inside in more immediate danger; and putting on 50-plus pounds of protective gear to perform arduous physical labor under extreme conditions are all activities that are unnatural. Most of society involves itself in activities that do not involve endangering their lives or require extending themselves to the very limits of human endurance. The consequence of casual leadership in the civilian world is thus less likely to endanger the lives of those being led; however, the exact opposite is true of the fire service (fig. 11–3).

Fig. 11–3. The "Miracle on the Hudson" was a once-in-a-career incident where everything needed to go right. Fortunately, it did. Full-contact leadership makes for a fire, rescue, and emergency force that can respond to and successfully mitigate even the most unthinkable situations. Full-contact leadership is the antidote to the unthinkable situation. (Courtesy Ron Jeffers)

Sensual leadership is reactive, from the hip, and rash. Sensual leadership creates more problems than it solves. Sensual leadership often follows the confused and tragedy-rich path of "Shoot! Ready! Aim!"

Doing what feels good on the fireground is commonly known as freelancing. It is universally understood that freelancing at a fire, emergency, or rescue operation places other operators and the mission at risk. Freelancing does not just happen; rather, it is allowed. Freelancing is a failure of leadership, with its root in the pre-emergency environments. Thus, it is a top-down creation. What a department permits, it promotes.

The curative prescription for a case of freelancing is diligent, competent, and interventionist leadership (fig. 11–4). Even with the most conscientious oversight, all freelancing may not be eliminated. However, with such oversight, freelancing will become an exception instead of the rule.

A casual approach to leading firefighters is not too hard to recognize. Again, you know it when you see it!

Fig. 11–4. Accountability is a discipline that must be assigned during hard-environment operations. A full-contact fire ground commander uses the command post as an organizational manifold maintaining a visual record of operational activities. (Courtesy Ron Jeffers)

chapter **12**

ABSENTEE AND AMBUSH LEADERSHIP

You learn from negative leadership more than from positive leadership because you learn how not to do it therefore, you learn how to do it.

—Herbert Norman Schwarzkopf Jr.

Absentee Leadership

Absentee. *A person who is not present in a usual or expected place; a person who is absent.*

A fire service leader is defined by the perceptions of those who are led. An absentee leader is unable or unwilling or fails to project the authority of an assigned command position.

Absentee leaders are easily recognized. Excuses, blaming, and specious reasoning are the stock-in-trade of absentee leaders. Their instructions are vague, interactions with them are confusing, and their actions and comments are unproductive or even counterproductive. Much of the time, absentee leadership is expressed through nonverbal communication and inaction. Hoping things will be okay and wishing things had gone better are the working philosophies of the absentee leader.

Absentee leadership is always working to catch up. However, catching up and leading cannot exist in the same space. Absentee leadership is prohibitively expensive and contagious. It is a black hole into which money, valuable time, positive energy, and employee motivation disappear. Under absentee leadership, the safety and well-being of those led and those served are routinely jeopardized, and the goals of the organization go unmet.

Proximity is not always the measure for absentee leadership. The most dramatic and harmful effects of absentee leadership often occur when the absentee leader is physically present.

A hallmark of absentee leadership is having an *under-the-rug filing system*. Sweeping issues under the rug is a variety of misfeasance, nonfeasance, or malfeasance that routinely occurs; nevertheless, through absentee-minded logic, issues that should be addressed instead are filed under "it's not that big a deal," "more trouble than it's worth," or, worst of all, "it's not my job." Examples of statements that reflect under-the-rug filing include:

- "He's only been late a couple of times this month."
- "Just sign off on the inventory. It won't make a difference."
- "She doesn't have her gloves on, but she doesn't work for me, so I don't have to say anything."
- "Who's watching the watchers?"

Absentee leaders are not made; rather, they are permitted. Absentee leadership is directly attributable to an organization's failure to hold officers, at every level, accountable and responsible. Absentee leadership finds root in an organization that does not continually educate and reinforce the roles and responsibilities of leadership positions. The only cure for absentee leadership is direct intervention up and down the chain of command. Department-wide vigilance is required in order to move absentee leadership out of its comfort zone, toward a more present and effective style of leadership. Reeducation and large doses of accountability are necessary to make present an absentee leader.

Absentee leadership never follows a bottom-up path; it is always a top-down progression. Ultimately, the chief of department is accountable for any and all leadership failures, minor or major.

Just as dogs tend to look like their owners, a fire department resembles its fire chief, a division resembles its division chief, a battalion resembles its battalion chief, and a company looks exactly like its company commander. If the chief of department is regularly absent from day-to-day operations—for example, is distant, hands off, reticent to intervene, or prone to errors or omissions of convenience or takes a myopic view of the department—then all of the officers under his or her command will follow this lead. By contrast, if a chief of department is present, hypervigilant, discriminatingly hands on, and known to intervene whenever necessary, command and junior officers will respond and lead accordingly (fig. 12–1).

Fig. 12–1. Full-contact Chiefing requires the chief to be truly present on a regular basis. The Chief's presence should not be the equivalent of a Big Foot sighting. Note that all officers shown here are in full PPE. Want to set a good hard-environment example? Start by wearing your gear.

You Know It When You See It

What you do has far greater impact than what you say.
—Stephen R. Covey

As part of a commitment to the enhanced safety and effectiveness of all operators during fire and emergency ground activities, the department issued a state-of-the-art portable radio for use by every on-duty operator. Every member of the department would be required, by SOP, to carry and monitor assigned portable radios during all fires and emergencies, as well as any activities conducted outside of company quarters. The SOP specifies when, how, and where all portable radios are to be carried.

Each of these radios was outfitted with a collar microphone, a leather case, and a leather shoulder strap that is collar-mic ready. The chief of department personally briefed all line and staff chief officers on the department's commitment to enhanced fire and emergency ground safety. The chief of department emphasized that improving fire ground communication capabilities was the goal central to the safety program.

The division and battalion commanders were provided with a training schedule. They were informed that the training division had made arrangements for a manufacturer's representative to assist in the initial training and familiarization. The chief closed each briefing with a clear statement that he expected all department-ordered radio procedures to be universally enforced. No exceptions!

Everyone knew that the chief of department would occasionally ride in on alarms to observe the department at work. It was during one of these ride-ins that the chief observed Engine 4's pump operator standing at the pump panel with an uncased radio sticking out of his back pocket, collar mic clipped to the antenna. Upon returning to headquarters, the chief of department informed fire dispatch to have the division and battalion chiefs report to his office. The chief of department took the following actions:

- The division chief and all battalion commanders were ordered to personally conduct an immediate inspection of all radios in all companies and battalions.

- The division chief was ordered to reinforce the requirements of the SOP and to make right any and all compliance issues. This action ensured the in-service and ready status of all on-duty members.
- When the inspection was complete, the division chief, the battalion chief, and Engine 4's company commander were ordered to report back to the chief of department.
- The chief of department reminded the assembled officers that they would be held accountable for any failure to comply with radio SOPs by members under their command.
- The chief of department informed the division, battalion, and company commanders that each officer was to complete a failure-to-supervise report, to be forwarded to his office by the end of the day.
- The chief of department made it clear that no report would be required of the offending firefighter.
- After reviewing the reports, the chief of department met again with the division, battalion, and company commanders.
- The chief of department informed all three officers that they were guilty of a failure to supervise.
- Because the failure had to do with life safety of personnel, the division chief, battalion chief, and captain were further informed that each of their reports would stand as an oral reprimand and be maintained in their individual personnel files.

The chief of department held the chain of supervising officers responsible for the firefighter's failure to comply. The firefighter's infraction was attributed to absentee leadership at the level of the supervising officers, which was carried up the chain of command. Because the unprotected radio sticking out of the firefighter's pocket was not the problem, but rather a symptom of absentee leadership, disciplining the firefighter would not address the root issue; even though negative discipline might make the officer corps feel like the failure to comply was addressed, such a response would be counterproductive and, more important, misdirected because the fault was instead with the division, battalion, and company commanders. A preeminent responsibility of a chief of department is to hold supervising and command-level officers accountable for the actions of those under them.

In this case, the chief of department issued an order for follow-up training on radio SOPs to be conducted department-wide, to begin immediately. The chief of department also scheduled a second series of meetings with all line chiefs and officers. At these meetings, the chief of department reinforced the mandate that all radio SOPs were to be followed and emphasized that the role and responsibility of each officer was to ensure compliance with any and all safety procedures. Officers were informed that they would be held directly accountable for ensuring that all operational procedures were followed.

Safety of operating personnel is the responsibility of every fire department. The chief of department's experience suggested that if one company was not following procedure, then the possibility existed that other companies were not following procedure either. With that in mind, department-wide follow-up training was conducted.

The chief of department believed that every order, policy, or procedure issued was important, and if it was important it was vital that compliance be universal. The mandate had to be made explicit or risked being missed. Thus, if any matter was important enough to require issuing an order, it was imperative for that order to be enforced. If the chief failed to ensure compliance, then the message that would be perceived by the department would be that compliance was optional (fig. 12–2).

An ongoing, comprehensive, leadership development program is essential to creating and maintaining a premier, 21st-century fire, rescue, and emergency leadership corps. If leadership is state of the art, then the department will be state of the art.

Positive and negative discipline are both integral parts of a leadership development program. Positive discipline and proactive development is always preferable to postincident corrective action. However, incident-specific and development-focused corrective action must never be avoided; when intervention is required, failing to intervene tacitly condones noncompliance. Punitive action should always be designed to teach and should be focused on changing the officer's behavior. Recalibration—of misperceptions and misaligned expectations—is the intended goal of all discipline.

chapter 12 | Absentee and Ambush Leadership 123

Fig. 12–2. Full-contact leaders always show up. The company officer translates policies and procedures into action. (Courtesy Ron Jeffers)

Officer development need not be confined to formal classroom instruction, lectures, and seminars. Mentoring is a learning and teaching tool that should be encouraged. Officers of rank should be schooled in the art of mentoring their subordinates. Mentoring as a form of developmental training resembles a floating crap game; that is, it holds the advantage of street-level immediacy and reinforces personal and professional bonds. Mentoring creates mentors.

Ambush Leadership

A boss creates fear. A leader creates confidence. A boss knows it all. A leader asks questions. A boss makes work drudgery. A leader makes work interesting. A boss is interested in himself. A leader is interested in the group.

—Russell H. Ewing

Dabbler. To work at anything in an irregular or superficial manner; one who dabbles; one not deeply engaged in or concerned with something (fig. 12–3). **Synonyms:** amateur, dilettante, hobbyist, nonexpert, nonprofessional, tinkerer.

Fig. 12–3. The fire scene pictured above is definitely a no-dabbling zone. Full-contact leadership makes its bones at highly complex, fast-moving, and strategically challenging incidents. (Courtesy Ron Jeffers)

Impossible as it may seem, there is another form of leadership that rivals and often surpasses absentee leadership as the most counterproductive and disruptive form of fire service leadership. The ambush form of leadership is harder to recognize because the ambush leader will get busy leading the moment that a superior officer looks over the ambush leader's shoulder. Hallmarks of ambush leadership include inconsistency, putting things off indefinitely, lack of or poor planning, hasty decision-making, excuse making, blame shifting, unmet or partially met goals and assignments, feigned forgetfulness, carelessness, rushing, panic, and confusion—and justifiably disgruntled subordinates.

An ambush leader is an absentee leader who shows up unexpectedly with a whirlwind of surprise orders that tend to be confusing and difficult to comply with. Ambush orders are sprayed around often purposelessly and are often issued just for the sake of giving an order.

The personnel assigned to the command of an ambush leader never know what will happen next. When there is no planning, there can be no scheduling. The absence of planning and scheduling create unnecessary stress, as well as dissatisfaction and resentment among subordinates. Scheduling is an extremely valuable leadership tool. Scheduling is not only a management tool, it is a statement of respect for the time and energy that subordinates are asked to contribute to the department. The importance of good scheduling cannot be overstated. A key to full-contact leadership is clearly stating expectations with regard to duties, and a schedule is a declaration of expectations. Scheduling reduces stress by telling personnel what is going to happen on a given hour, day, month, year, and so on (fig. 12–4).

Scheduling can take the form of a published or posted chart, or assigned duties and activities may simply be relayed verbally. In either case, all scheduling should include a due date.

Including subordinates in the scheduling process allows subordinates control over their work environment. It has been shown that having a stake in the work environment is one of the most important factors that determines employee job satisfaction.

As with absentee leadership, ambush leadership never follows a bottom-up path; it is always a top-down progression. Vigilant and engaged supervision, along with communication of unambiguous expectations, must be routine. Standards for compliance must be set and consistently upheld.

North Hudson Regional Fire & Rescue
Monthly Activity Schedule

Date: 01/01/15 Thursday	Date: 01/05/15 Monday	Date: 01/09/15 Friday	Date: 01/13/15 Tuesday
Personnel not available for training Rivera Lombardo J Ramos Krieger Peralta Bodega	Personnel not available for training	Personnel not available for training Rivera	Personnel not available for training
Activities: Safety Matters	Activities: Haz-Mat Ops Awareness, Right To Know, Emergency Response Guidebook, Batt 1 @ 16th St. 09:00 Battalion Meeting CO Roundtable Harasment / Discrimination	Activities: SCBA on Air w/ TIC & Pak Tracker Driver Training	Activities: Ground Ladders w/ Knots and Tool Hoisting Driver training
Date: 01/17/15 Saturday	Date: 01/21/15 Wednesday	Date: 01/25/15 Sunday	Date: 01/29/15 Thursday
Personnel not available for training Rivera	Personnel not available for training	Personnel not available for training Rivera	Personnel not available for training
Activities: Light Rail Familiarization TBA Company Officer Smoke Ch 1 & 2	Activities: Petzl Platform & Grade TBA Safety & Survival Ch 1 & 2	Activities: General Orders Sec 5 Rules & Regs Art I & II	Activities: Hose Stretch Dry w/ Forcible Entry

Division __1__   Battalion __1__   Month / Year: __January–2015__

Battalion Commander   **Mark Lorenz**   Digitally signed by Mark Lorenz
DN: cn=Mark Lorenz, o=NHRFR, ou=198, email=mlorenz@nhrfr.us, c=US
Date: 2014.12.28 14:31:51 -05'00'

Fig. 12–4. North Hudson Regional Fire & Rescue monthly training and activity schedule

chapter 12 | Absentee and Ambush Leadership

You Know It When You See It: Something Real That Might Have Happened

Captain Neutron and Captain Proton had tied for the top spot on the battalion chief's test. There were two assignments open. One was a lusted-after operations posting, and the other, a less attractive administrative job. Not surprisingly, both Neutron and Proton were hot for the operations position. Whatever their unfathomable reasoning, the assignment gods placed newly minted Battalion Chief Proton in the coveted operations position, and the shiny new Battalion Chief Neutron was relegated to the less prestigious job of training and facilities chief.

Because he had a few more years on the job than Proton, Neutron believed that he was entitled to the operations post. Neutron felt resentful about being passed over for his dream job.

Neutron's resentment spilled over into the way he approached his duties and delegated assignments. It was clear to everyone within Battalion Chief Neutron's sphere of influence that any assignment Neutron was given was carried out grudgingly and with little regard for results. Neutron would never do anything unless it was specifically mandated and monitored by his superior.

As a company commander, Neutron was a well-known master of "make-work" projects. On the occasions when Neutron was forced to or felt he needed to do some "chiefing-it-up," he would show up unannounced at a fire station immediately after shift change with a list of work that he wanted completed that day. Naturally, this drove the company officers up the wall and created considerable tension between the line and staff officers. As often happens, the company officers would sigh, shake their heads, and do the best they could to carry out Neutron's make-work directives.

Besides being a disruptive force for the line personnel, Neutron's helter-skelter management style was not bringing about the results that his boss, Deputy Chief Nucleus, demanded. Deputy Chief Nucleus expressed his displeasure during a morning briefing.

In response to the unwanted scrutiny focused on his performance by Deputy Chief Nucleus, Battalion Chief Neutron went into overdrive. The department maintenance schedule called for apparatus and station inspection to be conducted by Neutron's boss, namely Deputy Chief Nucleus. Neutron knew that Nucleus was a stickler for washed, cleaned, and waxed fire apparatus. Deputy Chief Nucleus had always been very specific regarding his abhorrence of the accumulations of refuse and storage of any machine fluids such as motor oil and hydraulic fluid. However, with the dates of inspection quick approaching, Neutron was confronted with two harsh realities: First, he had not ordered or procured the special detergent used to wash the vehicles; and second, a half-full 55-gallon drum of waste oil was sitting in the alley between Ladder 22's quarters and the five-story tenement building that stood next door.

Panicked and under tons of self-induced stress, Neutron came up with what the entire fire department had come to label as another one of his "Plan of the Day" ideas. Neutron's answer to the detergent problem was to supply each unit with a substitute cleaning agent. Unfortunately, it happened that the cleaner he distributed was in fact an abrasive cleansing powder. When the companies received the abrasive cleanser, they immediately pointed out that scouring powder would not be appropriate for the job. Neutron would not hear a word of it and ordered the companies to have all units washed and waxed within the hour.

Meanwhile, Battalion Chief Neutron sped off to deal with the half-full barrel of waste oil at Ladder 22. Ladder 22 was out of quarters on a squad call, so Neutron, responding to his panicked sense of urgency, devised a do-it-himself action plan: he would move the barrel out to the curb himself, then call the department of public works to pick up and dispose of the waste oil. Neutron took a hand truck from the station storage room and went to move the offending barrel of oil. Although the alley presented a few topographical hurdles that caused waste oil to slop over onto the concrete path between the adjacent tenement and the firehouse, the battalion chief successfully moved the 55-gallon drum to the curb in front of Ladder 22's quarters.

Battalion Chief Neutron went about returning the hand truck to the station storage room. However, in his haste, he drove the hand truck through the puddles of oil in the alley. Consequently, a trail of black and slippery tracks and shoe silhouettes marked his path from the alley door through the kitchen to the station phone.

Then, Battalion Chief Neutron called the department of public works. Neutron informed the public works director that a barrel of waste oil was waiting to be picked up on the curb in front of Ladder 22's quarters. The director returned the favor by informing Neutron that waste oil had to be disposed of by a certified waste removal firm and that it would take a day or more to schedule a removal.

Just then Ladder 22 began backing into quarters while Deputy Chief Nucleus pulled up to the firehouse. With the phone still in his hand and black goo on his shoes, Neutron looked up and stared open mouthed at the ladder truck, whose once-shiny fire-red paint job had been scoured down to a dull and funky pink pastel.

Even fish can recognize bad leadership.
—anonymous

He who cannot be a good follower cannot be a good leader.
—Aristotle

chapter 13
MORALE AND MOTIVATION

The best morale exists when you never hear the word mentioned. When you hear a lot of talk about it, it's usually lousy.
— Dwight David Eisenhower

Morale is self-esteem in action.
— Avery Weisman

Morale still seems reasonably high and, while the desertion rate has risen, it is still limited to those who can walk.
— Woody Allen

Ability is what you're capable of doing. Motivation determines what you do. Attitude determines how well you do it.
— Lou Holtz

Motivation is the art of getting people to do what you want them to do because they want to do it.
— Dwight David Eisenhower

Do the right thing it; will gratify some people and astonish the rest.
— Mark Twain

The best way to appreciate your job is to imagine yourself without one.
— Oscar Wilde

Morale and attitude are fundamentals to success.
— Bud Wilkinson

Stand up for what you believe in, even if it means standing alone.

—anonymous

If you don't like something, change it. If you can't change it, change the way you think about it.

—anonymous

Prologue

When I was a captain, I attended a fire science course conducted by Battalion Chief Frank Constantinople of the Jersey City Fire Department. Once the class was concluded, I made my way to the front of the room to thank Battalion Chief Constantinople. Wanting to talk a little longer, the battalion chief asked me to wait until he was finished getting his stuff together and speaking with a few other students who were vying for his attention.

While I was waiting for the instructor, I was approached by two firefighters from the Jersey City Fire Department, who quickly provided me with the 411 on Battalion Chief Frank Constantinople. To paraphrase their assessment, Constantinople was a first-in/last-out, my-way-or-the-highway type of fire officer from the school of hard knocks—unflinching, straight to the point, wicked smart, an ex-marine, and tough as a Halligan tool. I was further informed that Frank possessed an encyclopedic and savantlike recall of all things fire department and fire science related. There were two other things that my newfound friends made certain I understood: First, Frank was not at all easy to get along with and did not suffer fools or foolishness readily; and second, Frank would someday, without a doubt, be the chief of the Jersey City Fire Department.

Coincidentally (or not), the similarities to Chief Flood are undeniable!

—Avillo

chapter 13 | Morale and Motivation

After Battalion Chief Constantinople finished with the other students, he made his way over to where I was waiting and without fanfare asked if I would be interested in starting a study group with him. I confess that I was surprised and intimidated. Of course I said yes—I jumped at the chance to study with such a heavy hitter, himself the epitome of a serious student of the fire, emergency, and rescue business. So began my five-year, ad hoc, home-schooling equivalent to an advanced degree in fire science.

During one study session, I held a long-winded, soapbox dissertation on the subject of morale. When finished, I stood, hands on hips, basking in the glow of my prodigious intellectual prowess, expecting to receive congratulations—even a call informing me that I had been elected president and resident genius of the international fire service debate team—but things didn't quite go that way. Battalion Chief Constantinople had listened as politely as he could but eventually had enough of my naiveté. He said, "Kid, you wanna know what morale really is? Morale is just a bunch of feelings. And I'll tell you another thing. The only time anyone hears about morale is when something is going to be bad for morale. Nobody has ever come up to me and said, "Gee whiz, Chief, morale is really good today, thanks a lot." Initially, I was crushed, speechless, confused, and indignant. However, as usual, he was right: "Morale is a just a bunch of feelings."

Few work related-issues are more difficult to deal with than people's feelings (morale and motivation). It is doubly hard in a paramilitary organization composed of tight-knit units of men and women who perform myriad great and small mercies and, by choice, enter the eye of the storm to conduct heroic and courageous work in dangerous and hazard-rich environments.

Dealing successfully with morale and motivation issues requires a real-deal top-down, department-wide commitment, The "real and the deal" must start inside the office of the chief of department. All line and staff personnel, every command and company-level officer, and every firefighter in the department must be fully invested. The department must be prepared to fully engage with and educate all command and junior-level officers about the investment versus return on morale and motivation. The department must encourage and support innovative, creative leadership throughout the entire organization. Most important, even under the most enlightened leadership, morale and motivation will never always be high, and high morale will never be universal. Leaders and organizations must learn to expect and deal with periods and pockets of low morale and motivation. There is always an ambient level of background noise consisting of griping, beefing, and complaining. This is how individuals, units and the organization off-gas. Supervisors must be able to discern normal day-to-day griping from true grievances.

While a full discussion of grievance procedures is beyond the scope of this chapter, it bears explicit mention that all grievances, reports of harassment, and reports of unsafe or criminal behavior must be investigated and properly dealt with. Whether reported or observed, any situation that is or may be a misfeasance, nonfeasance, or malfeasance must receive immediate action and attention. All supervisory personnel need to keep an ear to the ground and an eye on the ball. If conditions do not feel or sound right, ask or investigate—find out what is really going on. Do not ignore any work-related situation that makes you uncomfortable.

Remember, the word "supervisor" comes from the Latin for extra vision. Thus, supervisors are supposed to go around looking at stuff, taking an extra-long look at all things under their supervision

Supervisors are often confronted with what I call *noncomplaint complaints*. These are frustration- or disappointment-driven gripes. Noncomplaint complaints are often redundant or specious, and—basically baloney. Filling out forms for these complaints is a way of off-gassing. However, reviewing these forms feels like a waste of time and is annoying, and it is easy to offhandedly dismiss the complainer and the complaint. Nevertheless, full-contact leadership requires the supervisor to separate the complaint from the person doing the complaining. Although a noncomplaint complaint is often a form of attention getting and the gripe might itself be a noncomplaint, the person bringing forth the issue can never be treated as a nonperson.

I have a bawdy story (mildly sanitized here) on hand that I tell (or inflict) in response to sincere gripes. My theory regarding routine complaints is that if I were to issue each and every firefighter and officer on every shift a brand-new wristwatch at 0900 hours, I would be in receipt of a grievance by 1100 hours claiming harassment and discrimination because I had neglected to consider that some members were comfortable only with pocket watches. At 1105 hours, I would be put on notice that the Occupational Safety and Health Administration would be inspecting the wrists of all firefighters and officers to ensure that no wrist-related injuries had occurred and that the department would be required to provide documentation certifying that none of the watch bands contained lead or other hazardous metals.

Morale is ethereal. Whether it is high or low is subject to emotional and mental posturing—and often seems like an emotional coin toss. Morale is a state of mind comprising the spirits of individuals, groups (formal and informal), and even an entire organization. High morale manifests as cheerfulness, self-discipline, confidence, esprit de corps (fig. 13–1), and willingness to perform assigned tasks.

Fig. 13–1. Esprit de corps—the sense of team and oneness of purpose—builds confidence and imparts a sense of group identity that leads to contagious motivation, high morale, and accomplishment.

Low morale often creates competing factions. Low morale can be exhibited in the way a person, a group (formal or informal), or an entire organization responds in the face of opposition or hardship (or both). Employee morale represents the spirit, or tone, of an organization. Employee morale affects individuals, the work environment, and all formal and informal relationship constructs within an organization.

Employee morale is based on work-related issues and circumstances. Good employee morale means employees are generally happy with their jobs. Low morale is an indicator that something in the work environment is off-kilter. Whereas high morale breeds a productive and efficient workforce, low morale leads to absenteeism, unproductive workers, indifference, and decreased motivation. Quality of output and performance suffer when morale is low.

The Hawthorne Experiments

Among the most interesting and informative studies of morale and motivation are the Hawthorne studies conducted by George Elton Mayo, a management theorist and professor of industrial research at the Harvard School of Business. From 1927 to 1932, Mayo conducted experiments on employee motivation and productivity at the Hawthorne Works factory complex of the Western Electric Company in Cicero, Illinois, near Chicago. Mayo's research made him one of the most famous names in management history.

Mayo's team conducted experiments throughout the Hawthorne factory for several years, but it is his study of a group of six employees who worked on a relay assembly line that would forever change how employers and supervisors viewed the paradigm of motivation and productivity and management/employee relations. The experiment took place over five years, and during the study period, a single researcher was assigned to monitor and interact with the relay assembly operators throughout their shift. The researcher would take notes and record everything the employees did and regularly informed the workers as to what was being recorded and how the experiment was progressing. The relay assembly operators were routinely consulted and asked questions about their work. The workers were solicited for input and feedback regarding their work and their feelings about the study.

Every few weeks, a series of closely controlled changes were implemented. Each set of changes would last one to three months. Normal working conditions for the relay assembly workers were a 48-hour workweek including Saturdays and 8-hour workdays from 8 a.m. to 5 p.m.; workdays did not include any rest periods, except an unpaid break for lunch. Under these conditions, the relay assembly workers produced

2,400 units per week. All changes were measured against the worker's normal working conditions, and 2,400 units per week was used as the baseline for productivity calculations.

The study began by adding a short break from work and reducing the number of hours worked per day. The result was that output increased. Then, the assembly line was modified to a more efficient "piecework" method, which is defined as work done "by the piece" and paid for at a set rate per unit. Workers were paid according to the number of units produced. As a result, output increased again. The next series of changes provided the relay workers with extended break periods, once again with positive results. In fact, output rose sharply.

The next change was to cut the time of breaks in half and increase the number of rest breaks from two to six breaks a day. At this point, the employees complained that their work rhythm was broken by the frequent number of breaks. As a result, output dropped slightly. Subsequently, the number of rest periods was returned to the previous routine of two a day. With this change, though, the first break period would include a hot meal supplied by the company free of charge. Output went back up as a result. Within a week, the workday was shortened by 30 minutes. The workday ended at 4:30 p.m. instead of 5:00 p.m. Once again, output went up.

When the study came to an end, all upgrades and improvements were taken away. Work conditions returned to the exact conditions that existed at the beginning of the experiment for a period of 12 weeks. The result at the end of the study (with conditions identical to those at the beginning) was that output was the highest ever recorded, with each employee producing on average 3,000 relay units a week.

Hawthorne experiment findings

At the end of the five-year period, the working conditions for the six relay assembly workers reverted back to the exact conditions that existed before the experiment began. Unexpectedly, the workers' morale and productivity rose to levels higher than ever recorded before including during the experiments. The combination of results during and after the experiment—namely, the increase in the worker's productivity when they were returned to their original working conditions—suggested to Mayo that workers were motivated more by psychological than physical working conditions.

Hawthorne experiment conclusions

The commonly held belief that people work purely for money and to make a living seemed to be deeply flawed. A person's work was much more than a way to earn money and make a living. Work was first and foremost a group activity, where people had an effect on other people. That is, individual behaviors affected the behavior of co-workers, and this holds true regardless whether the parties are colleagues, managers, or observers.

Morale and productivity are affected not so much by the conditions in which people work but by the recognition and attention they receive. The rises in productivity documented in the relay assembly room experiment were achieved because the employees were working under the interested eye of observers. Increased productivity was not the result of improved work conditions. Productivity improved because workers were allowed input into their work environment and were made to feel that they were important and valued members of the company.

Mayo's study concluded that workers were motivated by more than self-interest alone. The following factors also had an impact:

- *Psychological contract.* There is an unwritten understanding between the worker and the employer regarding what is expected.
- *Interest in workers.* Worker motivation can be increased when an interest in them is evident. This might be from within or outside the organization; for instance, Mayo classified studying the workers (throughout the experiments) as showing an interest.
- *Work conceived as a group activity.* Teamwork can increase worker motivation because it allows people to form strong working relationships. Teamwork increases trust between workers. Work groups are created formally by the employer, but workers create informal groups within the formal work group. It should be acknowledged that both informal and formal groups exist. Informal groups have the power to influence workers' habits and attitudes. The informal group can impose greater influence and have a greater impact on the productivity, success, and cohesiveness of the formal group than any other factor. Appropriate, well-informed, and well-intentioned manipulation of informal and formal work groups can and should be used to create harmony and a productive work environment (fig. 13–2).

Fig. 13–2. The Knot Gauntlet competition was held over a three-month period and took on a life of its own. Spying, rumor, bravado, and chest-thumping all became part of this motivation-in-action competition. Each battalion's champion squared off in the final round to win this nifty, no-expense-spared plaque. The Knot Gauntlet is an example of creative, full-contact leadership that reinforces rope rescue skills, team-building, vital group image, and self-realization while at the same time creating an environment of mentoring, learning, and skill advancement.

- *Social aspect of work.* Workers are motivated by the social aspect of work. The assembly line workers socialized during and outside work. This socializing was a major contributing factor for the noticeable increase in motivation.
- *Worker recognition.* Workers are motivated by recognition, security, and a sense of belonging.
- *Top-down and bottom-up communication.* The communication between workers and management influences workers' morale and productivity. Workers are motivated through a good working relationship with management.

The traditional view of how to motivate employees is to offer monetary rewards (pay increases, bonuses, etc.) for work completion. However, the Hawthorne experiments suggest that motivation is more complicated. Advocates of the *Hawthorne effect* assert that the Hawthorne experiment results show that motivation can be improved through improving working relationships and social interaction.

Minneapolis Gas Company Employee Motivation Study

The Minneapolis Gas Company studied 31,000 men and 13,000 women over a period of 20 years (1945–1965). The goal of the study was to identify and determine what employees want most from their job. Surprisingly, factors such as pay, benefits, and working conditions were given low ratings by both men and women. Instead of money and working conditions, all employees rated job security as the most important job satisfaction factor. Following job security, employees identified promotion, advancement, the type of work being done, and pride in the organization to be the most important issues when it came to job satisfaction.

Morale, Motivation, and the Real World

When I first got on the "JOB," firefighters were still smoke eaters, the fire triangle was just the fire V, and rehab consisted of sitting down on the curb, taking your helmet off, and firing up a Lucky Strike. Morale was for sissies, and motivation had something to do with a faulty carburetor. All of the officer corps and the majority of the rank and file were of the "Greatest Generation"—particularly, World War II and Korean War veterans. They were a great bunch of firefighters—Lucky Strike–smokin', smoke-eatin', and with no time for sissies who would tell you to take your morale and put it where your motivation don't shine! By contrast, I am a child of the '60s—a tree-hugging, mantra-chanting, flower-power, transcendental-meditating, protest-marching, touchy-feely, hippie-dippie dude. Thus, with no disrespect to my early mentors, I will broach the subjects of morale and motivation.

chapter 13 | Morale and Motivation

A fire service leader is tasked to get work done through the efforts of others. To do this, an officer needs to motivate subordinates. Of course, motivation is much easier said than done. This is because motivation has to do with human beings.

When motivation is positive and effective, it becomes a process that can initiate behaviors allowing the group, unit, organization, or department to accomplish a mission or achieve a goal. Motivation is what causes individuals and groups to act.

Motivation is tricky; it is ethereal, a lot of work, and can at times be a big pain. Much of the time, motivation is looked upon as an unattainable, philosophical goal. One school of thought believes that motivation is something that sounds good in a speech but does not have practical application in the real world. In fact, management and leaders can have a direct and positive impact on employee morale.

Human beings are the most high-maintenance and overly complicated asset in any fire department. Motivation is hard because it involves the biological, emotional, social, and cognitive forces that drive human behavior. And when you put it that way, who in their right mind would want to tackle that beast? For a supervisor to get stuff done through the efforts of other humans, the supervisor must feed the beast a perfectly blended high-octane fuel; motivation is the high-octane fuel that moves the human beast to do the stuff that the supervisor wants done.

You cannot issue an order that mandates motivation. Motivation is not delivered and cannot be borrowed. Instead, motivation is individuated. No one can motivate anyone else. The only thing a leader can do is create an environment that brings the needs and aspirations of the individuals into line with the needs and mission of the organization.

The most valuable motivational tool in a leader's toolbox is creativity. As a matter of fact, if your leadership portfolio has only one file, make sure that file is the creativity file. I love the idea of creative leadership. A creative leader is inventive, positive, energetic, passionate, productive, intellectually curious, enthusiastic, and a prerequisite for full-contact leadership (fig. 13–3). Creative leaders are capable of initiating, instigating, and inciting positive change, effective action, and an ironclad dedication to mission. Through strength of vision, creative leaders are often able to inspire the people they work with to change expectations, perceptions, and motivations.

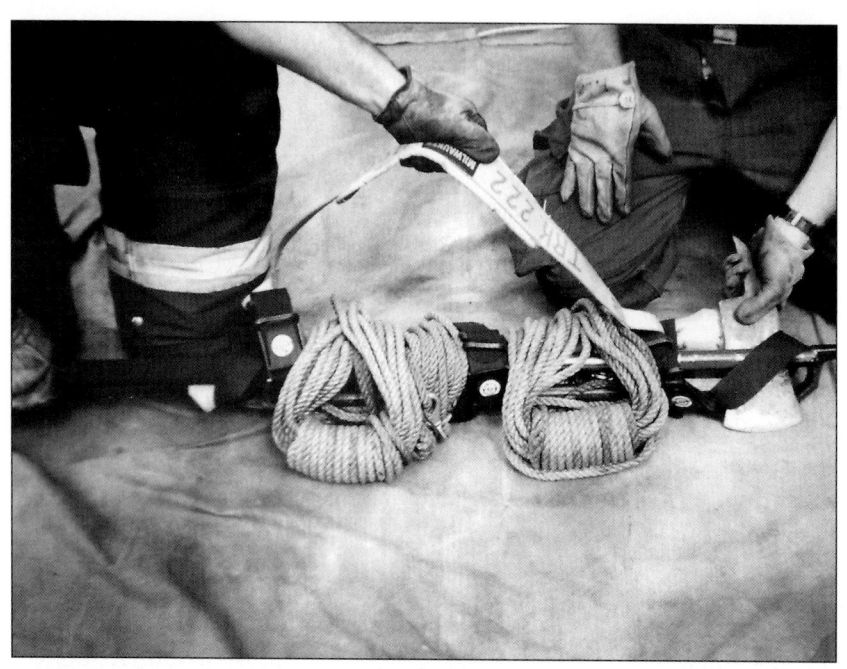

Fig. 13–3. The Roof Pack was put together by the members of Weehawken Fire Department's Truck 222, under the command of Captain Flood, to solve a firefighter-to-tool-ratio deficit. This is an example of a company motivated to adapt and make it work.

If you want to successfully create environments that are motivational, you had better be ready to be a creative leader. Creating a motivational environment is a conscious act. It is a contract, whether motivating a group or motivating oneself. Practicing creative leadership requires making a decision to accomplish a goal or to initiate or stop a behavior. Once a goal is identified, time, energy, and resources must be committed, and persevering through—and even embracing—adversity is the key component of a motivational contract.

Creative, full-contact leadership requires that a supervisor maintain the momentum, focus, enthusiasm, and concentration necessary to achieve desired results. This is where division of labor, supervision, and modifying actions are brought to bear on the process.

No leadership concept or philosophy stands alone. The purposeful application of management and leadership skills requires that a leader bring together some or all of the many supervisory disciplines.

The reader is invited—indeed, consider yourself challenged—to begin to synthesize, combine, meld, cross-reference, and mush together the concepts and information presented. A leader needs a leadership philosophy.

As Avillo Sees It: The Buy-In

An officer's job is to get subordinates to conduct the business of the department. It is best when people under your command can see the big picture—and buy into the big plan. Seeing the picture and buying into the plan is great but not always the reality.

It is the responsibility of a leader to motivate subordinates. A leader must show how cooperation and compliance with the requirements of the service benefit the individual, the unit, and the department. So how do you motivate them? You really can't—but you kind of can. People are motivated to act when acting satisfies or meets a personal need. Motivation is very much a personal choice.

It is a leader's job to create an environment where motivation can be bred and nurtured. A leader must demonstrate to subordinates how compliance and teamwork will bring the goals of the organization in line with the needs of the individual and the group (fig. 13–4).

Individuals and groups will extend better effort when they can see the value of the actions ordered. A leader needs to be both coach and counselor. At times cheerleading and parenting skills may be needed to move the individual or group along the desired path.

Give your people a say in what you are trying to accomplish. Invite their participation and ideas, giving them credit and more of a share in the creative process. That kind of leadership gives subordinates ownership in the process. Once they have ownership, people are more likely to buy into the program and infect others with their enthusiasm. There are times when a leader will have to work outside the box to achieve the desired results.

Fig. 13–4. Companies operating at this fire encountered exotic window protection. The metal window enclosures proved almost impossible to open. The next day provided an amusement park of opportunities to learn how to defeat these barriers. We started in the morning, worked together, shared ideas, learned a lot, and didn't stop until we came up with a plan to pass on to the rest of the department.

I witnessed this out-of-the-box-leadership firsthand in a training evolution at a warehouse. We were conducting confined-space training for a fire brigade at a plant, and there were some people who were less than interested in what we were trying to accomplish. Confined-space operations were new to the students, and the information may have seemed daunting. I believe that their disinterest resulted from a fear of the unknown more than anything else. The lead instructor came up with an extremely creative way to motivate them and ultimately brought them around by explaining that there were many parts to the job and that the need to work as a team was important. He gave reticent members the job of air monitoring and assigned them to operate the ventilation fan that would be put in the space.

The instructor explained that the assignment would be the first step in creating the vital life-support systems required. The task was success oriented. The instructor showed the timid participants that their actions could mean the difference between life and death of the person in the confined space. He dubbed this group the "A-Team," for air team. This bit of ownership in the job allowed the tentative operators to buy into the program. The instructor created a team building environment; the A-Team took a leadership role and found a sense of pride in what they were doing. When the A-Team members embraced their roles, it motivated others to step up and enthusiastically join the team.

The team came together as a result of the full-contact leadership demonstrated by the instructor. The instructor's understanding of the situation and a bit of creativity worked to reassure and motivate the students. In turn, their enthusiasm and motivation carried the whole training session to a higher level. If you, as an officer, can get those kinds of results from your personnel, then you have done your job as a supervisor.

chapter 14

DELEGATION

As Avillo Sees It

Chief Flood has spoken time and again about Prime Directives. As a division chief, my Prime Directive was to ensure the in-service and ready status of all personnel, all companies, and all battalions in my division.

My ability to meet the requirements of that Prime Directive depended upon my ability and willingness to delegate responsibilities to my company and command-level officers. Successful delegation required that I support my officers. My responsibility was to ensure that the resources needed were available. I was responsible for monitoring project progress. Whenever I delegated, I made myself available for consultation and to facilitate any action that requires decisions above the pay grades of the subordinates doing the work.

Introduction to Delegation

Fair and equitable division of labor is a concept that must be integrated into the delegation process. Supervisors commonly overdelegate work to the most motivated, competent, and willing subordinates. Those subordinates who have proven to be go-to workers are overly relied on. It is often said that 20% of the workforce does 80% of the work, and regardless whether this is truly the case, it is a common perception.

A primary function of a supervisor is to get work done through the efforts of others. It is when a leader and an organization focus solely on getting the work done, the jobs are inordinately assigned to the most productive subordinates. Such job dumping does nothing for employee development. Ignoring subordinate development in favor of getting the job done is a shortsighted and ineffective management philosophy. The organization, through its leaders, must do everything possible to integrate productivity with subordinate involvement and development.

When a leader does not have confidence in a subordinate, it is most often the result of a failure of leadership because subordinate competence is the responsibility of the supervisor. Ensuring the knowledge and competency levels of subordinates is a property of good leadership.

It is always easier and safer to delegate when a leader is confident in a subordinate's competence. Working closely with subordinates to develop knowledge and competency is harder and more precarious than dumping work on the 20% over and over. Failure by leadership to develop the knowledge and competency of the 80% is the most common reason for job dumping.

Sometimes a leader will be afraid to delegate for fear of being exposed as a weak leader owing to personal shortcomings or supervisory failures. Fear of losing control over subordinates is another obstacle to delegation. The truth is that delegation makes a leader stronger, more powerful, and more effective.

Some officers tend to think that a job can be done better when they do the work themselves. Falling back on an "I can do it better myself" attitude is one of the most common reasons leaders fail to delegate, ergo, fail to lead.

Developing subordinates into officers is one of the biggest responsibilities a supervisor has. Subordinate development depends on a leader's ability to trust people to accomplish the tasks they have been given. Training subordinates to a point where they are able to perform delegated duties without close supervision is where leaders earn their pay. The supervisor must develop trust and demonstrate confidence in those firefighters delegated to do the work.

An essential responsibility of an effective leader is to develop subordinates. The goal is to be able to confidently delegate tasks along with the authority and responsibility necessary to get the job done. All delegated assignments must consider subordinate competency and be success oriented. Delegation needs to follow a progression from less to more challenging, and delegating tasks progressively—from mildly challenging to very challenging—provides the supervisor with the opportunity to monitor a subordinate's progress. A known-to-unknown or a simple-to-complex progression also allows for points at which the leader can acknowledge work well done. Finally, a leader must remain available to the members doing delegated work—in other words, stay close but not too close.

More as Avillo Sees It

Delegation is the act of handing over control to someone else. Assigning responsibility to do a job or carry out a duty is delegation. Delegation assigns authority; however, ultimate accountability always remains with the person doing the delegating.

Delegation makes for a stronger leader, gives greater supervisory reach, stretches the span of control, and enables the accomplishment of much more work than could ever be done by a lone-wolf leader. Delegation empowers subordinates and allows the supervising officer to assess subordinate competency. Discovering and being able to take advantage of every individual's special skills, talents, and expertise is one of the major benefits derived from delegation of duties.

One of the most effective and efficient ways to develop officer proficiency is to delegate work whenever possible. Full-contact leaders make a conscious practice of integrating officer development into every activity. Officer development is an empowering process that builds leader confidence and hones leadership skills (fig. 14–1). Empowered company officers make powerful battalion chiefs—and having powerful battalion chiefs helped make me one hell of a division chief!

As a deputy chief, I took advantage of all delegation tools in my skill portfolio. Division of labor and specialization are foundational delegation concepts. All battalions in my division were assigned specific duties and areas of responsibility. The intent was to develop all battalion commanders into delegators and subject-matter experts. Each battalion delegator or subject-matter expert was responsible for developing lesson plans and drills to be conducted across the entire division. My subject-matter expert development program was possible because I was using and applying the full-contact leadership skills of delegation, fair division of labor, and specialization.

Fig. 14–1. Empowering both company commanders and subordinate chief officers creates a more effective and safe operating environment. This empowerment starts in the soft environment. (Courtesy Bill Menzel Photography)

For example, Battalion 1 was assigned to develop training on the types of foam extinguishing agents used by the department. Battalion 1 had designated one company officer as the foam guru. Captain Foam Guru was tasked with enlisting the cooperation of another company to assist in the development of lesson plans and in-service training sessions. In other words, I delegated to a battalion chief, the battalion chief delegated to a captain, and the captain fairly distributed the work. In this example, the battalion chief and the two captains specialized in foam as their subject matter. Voila! What we describe here is delegation, fair distribution of labor, and specialization. Pretty good, huh?!

A leader who tries to do everything will be chronically overwhelmed. Failing to effectively delegate will cripple and weaken any supervisor. Delegation is an exercise in command flexibility. Officers who cannot or do not delegate are inevitably paralyzed by their own inflexibility. The common denominator among effective and powerful full-contact leaders is the ability to judiciously delegate authority and responsibility to subordinates.

Remember that when subordinates succeed, leaders succeed—and when subordinates fail, it is a failure of supervision. When you delegate, you are delegating authority and responsibility but never accountability. The delegating supervisor is always accountable for the proper completion of the assignment, as well as the actions of subordinates.

As Avillo Sees It: The Gift of Imperfection

You will not be here forever. Train your people to take your place.

—Director Mike De Orio

Another of the many things I learned from Chief Flood was that when delegating, I needed to give myself the "gift of imperfection." Focusing on the how is a common trap when delegating. Delegation is an exercise in trust. No matter what you delegate, no one will do the job exactly as you would. The gift of imperfection comes with the warning to stay way out of the way. If you think your nose should be in the mix, maybe you shouldn't have delegated in the first place.

There is a glaring incongruity between delegating a task and then hovering over it like a mother hen. That is not delegation. Learn to be open-minded and accept a job well done even if the result was arrived at by a different path than you would have taken.

Failing to allow people to do things in their own way is not delegation. The entire philosophy behind delegation is to free the supervisor to do other important supervisory tasks. Delegation is designed to develop and empower subordinates while simultaneously building their self-confidence. When the boss is overly involved in the task, it kills motivation, causes resentment, and usually screws up the whole project.

> If a task is not done correctly, the first person to look at is yourself. Did you provide the time and resources necessary? Were you clear in your directions and your expectations? Were you available to be consulted? Did you teach your subordinates the exception rule: You do not have to consult with me unless there is an exception. An exception is anything that occurs that is totally unexpected or requires

> *a decision be made at a pay grade higher than yours. If there is an exception that I should be consulted on, I should be consulted without exception. If not carry on.*

Keeping your hands and nose out of the work conveys that you trust the abilities of your subordinate(s) to accomplish the delegated task.

The progress of delegated tasks should still be monitored. When delegating work to subordinates, a leader needs to establish benchmarks, such as:

- Set and agree on scheduling and deadlines.
- Ensure delegated task is properly resourced.
- Identify and schedule progress and review points.
- Build in enough flexibility to make adjustments when required.
- Carefully review all work when submitted.
- Recognize good work and commend people when deserved.
- Do not accept work that is substandard or shoddy.

Consider these recommendations prior to and during the delegation process. Delegation provides many opportunities for team building, motivation of personnel, acknowledgment of good work, and improvement of morale. Remember that delegation is primarily a subordinate development activity.

Guidelines for Effective Delegation

- Describe the overall plan.
- Explain why you have chosen to delegate the task(s).
- Inform any outside units, commands, or support facilities affected by the project.
- Facilitate any intradepartmental support.
- Include subordinates in the planning process.
- Allow subordinates to control the division of labor.
- Provide adequate support to ensure that all required resources are available.

- Be specific about expectations and scheduling.
- Clearly define time lines, who has authority, and areas of responsibility.
- Ensure that the level of authority is commensurate to the responsibility assigned.
- Dig deep. Delegate to the people who are closest to the work.
- Create and maintain fluid communications.
- Monitor, follow up, set benchmarks, and require progress reports as needed.
- Focus on results. Do not get involved with the how. Focus attention instead on what is accomplished.
- A leader cannot delegate away ultimate accountability. The buck stops with you!
- Create and project an environment that grows motivation and commitment.
- Stay close enough to be available for the exception but far enough away to not contaminate the project with "boss-think."

Final Word

Always be on the lookout for and take advantage of ways to provide opportunities for subordinates to take on challenging assignments. Share in their successes and guide them in fixing their failures. They may not even realize it, but what you are preparing them for is the next step in their fire service journey.

chapter 15

SETTING EXPECTATIONS

Being an effective officer and leader starts and ends with expectations. Before setting expectations for others, fire officers must first set expectations for themselves. Successful officers will develop and adhere to a code of conduct, reflecting both personal and professional values, that will serve as a leadership compass. Leaders need to be headed in the right direction because that is where the next right thing to do will be found. Personal and professional expectations should align with the needs of subordinates, superiors, and the department and its mission.

When setting expectations for subordinates, a full-contact leader will be clear and specific so as to be understood and acknowledged. Subordinates cannot be expected to follow directions if they are not told what needs to be done. Telling does not guarantee comprehension, so holding a question-and-answer session should be part of setting expectations: Ask the person or group if they understand what is required and if they have the ability to do what is expected; ask them to repeat what they understand to be your expectations; and ask if any information, support, or resources are required in order to meet those expectations.

Expectations are personal, professional, and organizational parameters and boundaries. Rules, regulations, policies, procedures, and schedules are expectations that have been formalized, and most expectations and directions for safe and effective firefighting operations are codified in department rules, regulations, SOPs, and policies. These tools are provided by the formal organization to guide and direct actions and decision-making, as well as to identify levels of accountability and responsibility.

Leaders are responsible for the personnel, equipment, apparatus, and facilities under their command. Commanders are accountable to the department. The department is accountable to the population served (fig. 15–1).

Fig. 15–1. Fire prevention activities create a contact point where the fire service educates and interacts with the customer. The public expects the fire department to provide a number of services, one of the most important of which is fire education. (Courtesy Ron Jeffers)

Officers are responsible for supporting the organization's mission and enforcing department expectations. When it comes to which formal expectation needs enforcing, there is no room for picking and choosing; fire service leaders must know, follow, and enforce all department mandates as well as meet department expectations themselves. Moreover, all officers must understand that department rules, orders, policies, and procedures are meant to be enforced fairly, consistently, and universally. To this end, fire officers must ensure that assigned personnel are educated and informed about all department expectations specific to their position.

The responsibility for all operational proficiency and compliance with requirements of the fire service is the job of the company officers, battalion chiefs, division commanders, and ultimately the chief of department. Every officer must understand that the fire department is a single unified fire, emergency, and rescue force. All fire operations are preplanned, cooperative, and coordinated activities. All fire operators must understand the symbiotic nature of the fire business; everyone has everyone else's back. To operate as an effective firefighting force, every firefighter and officer must meet the requirements and expectations that guarantee a consolidated effort.

Some expectations must be set in stone. Expectations that have to do with life safety and property conservation can never be compromised. Other expectations have a shelf life, and these need periodically to be reviewed, adjusted, amended, and sometimes discarded.

When setting expectations or giving orders, a full-contact leader will leave no room for ambiguity or misunderstandings. It is the responsibility of the officer to ensure compliance. Proper compliance is dependent on subordinates' understanding of what is expected.

Clarity of expectations is dependent on mutual understanding and agreement. Failure to comply and poor performance are often the result of poor communication and misunderstandings. Failure to come to agreement regarding the duties assigned or the expectations set will guarantee failure to comply (fig. 15–2). Firefighters want to do the right thing, and effective, engaged leadership is the recipe that guides firefighters to do the right thing.

Fig. 15–2. Department SOPs, preplans, and training activities set 95% of the expectations. Successful fire operations happen because the hardest work is done before the fire department arrives. (Courtesy Ron Jeffers)

Setting expectations is the key to consistency and fairness. Expectations should be set for all line and staff personnel and at every level of command, including the company, battalion, division, and department. To set expectations, officers must maintain fluid communication with subordinates and superiors. Subordinates have the right to know exactly what they can expect from their boss.

An officer does not have to justify every order given; however, *all orders must be justifiable*. Every order given and every expectation set must have purpose and must be reasonable, rational, logical, and in line with the requirements of the service. Firm, fair, and friendly are the respective tone, manner, and demeanor an officer should project when issuing orders or setting expectations.

As Avillo Sees It: A Mini Case Study

You are a new officer assigned to an engine company. This is your third tour of duty. It is 0745 hours. The tour of duty started at 0730 hours.

An alarm sounds for a reported smoke condition at a school. As you are donning your gear and checking the information from the computer-assisted dispatch system, you see two of your firefighters stroll out the kitchen. One of the firefighters has a cup of coffee in her hand, and the other is putting on his uniform shirt. The two firefighters casually make their way to the gear rack to get their PPE.

Are you getting uncomfortable yet? If not, you certainly should be.

Establishing Command Presence and Taking Command Action

I spend much of my time preparing my impromptu statements.

—Winston Churchill

If it ain't written, it ain't existing.

—Captain Kevin O'Driscoll

In this case, the hard truth is that you are not in as strong a position as you should be to take the errant firefighters to task. This is because you have already had two (which is one more than should be sufficient) tours of duty to explain how you intend to run the company and clearly set your expectations. What you consider to be the roles and responsibilities of the firefighters under your command should not remain classified information. Your expectations must be set early, often, and as needed. You are a fire officer, not a psychic; similarly, the firefighters under your command are equipped with PPE, not ESP.

It is your responsibility to fix casual attitudes and promptly reboot any erroneous perceptions held by firefighters under your command. The casual behavior regarding the PPE should have, quite rightly, made you uncomfortable. Perhaps these firefighters were allowed to operate casually under their former company commander, or maybe the officer you replaced tolerated a casual approach to the job. Either way, it doesn't matter. You are the boss now, and you are responsible for ensuring that your firefighters, all equipment, and apparatus are always in service and ready to respond. Life safety and property protection are two areas of responsibility that can never be compromised. There is no room for a casual approach to being ready to respond. The fire department's entire focus is life safety and property conservation. The reason citizens call the fire department is that they fear for their life or their property or both.

What should be done about your firefighters? Immediately upon return to quarters you should perform a quick attitude adjustment on the two firefighters. Do not wait—but also avoid falling into the quick-fix trap. Straightening the two firefighters out regarding the PPE is just a stopgap measure.

An officer must have both a vision and a command philosophy to guide action and decision-making. A leader needs to share and clarify both of these tenets with the assigned personnel. An officer should sit down, organize their thoughts, and come up with plans based on these two tenets. The belief held by many fire officers that they can hold forth extemporaneously on any and every command-related issue is unjustifiable. Expectations, vision, and command philosophy are integral to effective, productive, and successful leadership. Thus, the officer's plan, vision, and philosophy all need to be put to paper.

As Avillo Sees it

Proactive leadership makes things easier both for officers and for the people under their command. Its opposite, reactive leadership, is a come-from-behind, catch-up style of leadership. Reactive leadership is a fallback position that is called on only in the absence of proactive leadership.

Because there are situations and events that truly do come out of nowhere, there is a real need for reactive leadership in this business. No officer or department can predict every single event. However, proactive planning and action dramatically have been proved to reduce the need for such reactive leadership.

A good start for the new engine officer might resemble the following:

- Identify the duties and responsibilities that are required immediately upon reporting for duty. This includes immediately putting out PPE and ensuring that all self-contained breathing apparatus (SCBA) and personal alert safety system (PASS) units are in ready status.
- Perform a 360-degree check of the apparatus and compartments (firefighters and officer). The company commander should do the 360-degree check with the members of the company until confident that assigned firefighters can do the inspection on their own.
- Confer with firefighters and officer you are relieving regarding any changes or information that will have an impact on the in-service and ready status of the unit.
- Review the company journal.

Only after these duties are completed can any member of the crew have coffee or breakfast. The kitchen and coffee or breakfast provide an ideal place and time to brief the firefighters on the days' activities.

Activities listed above are a short list of expectations regarding a specific responsibility. The duties and expectations are straightforward—no surprises! Everyone is ready to roll, and everyone has an idea as to how the day will unfold (fig. 15–3).

Fig. 15–3. Companies en route to a fourth-alarm fire. Having everyone be ready to roll must be a department expectation. It is the single reason we exist. (Courtesy Ron Jeffers)

More As Avillo Sees It

Finis origine pendet. [The end depends on the beginning.]

—anonymous

Experience has taught that when I begin my day or any activity in the right way and move forward from that point in the right direction, things work out well in the end. *Primacy* is a learning concept whereby we learn best when we get it right the first time.

As a firefighter, I worked in a ladder company. My ladder company officer was Captain Ed Flood. Immediately upon reporting for duty, we were expected to do all the duties mentioned above. Once all duties were completed, we would be given our tool and riding assignments.

The company would review initial scene assignments and go over any other company-related matters.

Holding an early-morning "at-the-rig" meeting is a great way to set and reinforce expectations on a daily basis. It also is a chance to confirm the in-service and ready status of the apparatus and members. Meeting on a regular basis starts a dialog, opening lines of communication by members with the ultimate goal of getting everyone on the same page (fig. 15–4).

Fig. 15–4. Formal meeting or informal meeting? No matter what you call it, be aware that when it comes to issues of life safety or property protection, there is no such thing as informal firefighting. (Courtesy Al Pratts)

During our at-the-rig meetings, the entire ladder was checked out, every member was prepped, and all pertinent and necessary information was disseminated. In other words, the ladder company was in service and ready to roll. I was never in the dark and always knew what was what, who was who, and who was what. I was prepared, I was confident, and I knew what jobs I was responsible for.

Avillo Gets Promoted

There is nothing more permanent than change.

—Battalion Chief George Browne

I retired as the division commander of the 1st Platoon. Prior to my promotion to deputy chief, I worked as a battalion commander on the 3rd Platoon under Deputy Chief Flood. I felt comfortable on the 3rd Platoon. I was engaged and had invested time and energy building what I considered to be the best battalion on the entire department. I was promoted up the ranks and was in line for a division command assignment, and I was confident that I would eventually take over the 3rd Platoon once Deputy Chief Flood was promoted to chief of department. However, much to my dismay, Chief Flood transferred me and assigned me to be division chief on the 1st Platoon (fig. 15–5).

Fig. 15–5. The Fighting 1st—strong to the finish

At the time I felt entitled and was unhappy that I was being moved out of my comfort zone. Though I went to the 1st Platoon, I went kicking and screaming. In between my kicking and my screaming, Chief Flood told me that going to the 1st Platoon was the best thing for me at that time. The chief predicted that one day soon I would thank him. Chief Flood was right. I learned a lesson about comfort zones, for which I have thanked him many times since.

When I got to my new platoon, I knew some of the personnel, and I had a good relationship with the battalion chiefs. In fact, the 1st Platoon was the most senior shift on the department, but there had been no real leadership for quite a while. Instead of jumping in and making demands right away, I decided to give it a few weeks to a month, as an observation period before meeting with the shift and setting my expectations. During my observation period, we had many alarms, several room-and-contents fires, and one especially challenging multiple-alarm fire. I began to get a feel for who was who—and who wasn't.

It was at that first multiple-alarm fire that the firefighters and officers of the 1st Platoon were also able to take my measure. My command presence and competency in the hard environment helped the platoon members see what I was all about. That fire and the watch-and-see strategy enabled me to gain a great deal of trust from the officers and firefighters I had recently come to command.

During my observation period, I made sure to take notes and identify strengths and weaknesses of the platoon. By observing and documenting, I was able to develop a plan and a vision for my new command.

While I was in the wait-and-watch period, I had been conducting daily meetings with my battalion chiefs. My mantra at these early meetings was "Do more listening than talking; do more asking than telling." These early meetings created an environment that allowed my battalion commanders to understand that I wanted and needed them to be invested in a common command philosophy. The chief officers under my command would enable and implement the changes that I saw as necessary to move the platoon forward.

Once I had formalized my plan and understood what needed to be done, I met with the entire platoon. At this meeting, I laid out my expectations and explained my vision and command philosophy. The meeting set the tone and expectations for the future leadership of the platoon. I explained to the members of the 1st Platoon that I was there to provide support. I further explained to my firefighters and company officers that all of the 1st Platoon chiefs were there to help them do the next right thing. If the firefighters and captains were doing the right thing and meeting expectations, then the chief officers of the 1st Platoon would go to the wall for the people under their command.

My command philosophy recognizes that once a subordinate deviates from the path, there is a direct relationship between leadership support and the degree of the deviation.

The meeting described above took place more than ten years ago. In the interim, I have learned as much from the people under my command as they have hopefully learned from me. At the time of my retirement from NHRF&R, the expectations that were set at that initial meeting were still in place. Consistent compliance with and enforcement of expectations allowed the entire platoon to move forward in a unified and productive manner (fig. 15–6).

Fig. 15–6. High expectations. There was no shortage of supervision at the command post. (Courtesy Ron Jeffers)

My assignment to the 1st Platoon had given me an opportunity to grow personally. I was able to develop a platoon-level firefighting force in line with my command philosophy. Had I stayed on the 3rd Platoon, I doubt that I would have been as challenged. Instead, the assignment to the 1st Platoon ripped me out of my comfort zone, as a result of which

I got to put my stamp on a platoon. The assignment was one of the greatest favors Chief Flood could have done for me. I was permitted to cast my own shadow, and my shadow was good.

chapter 16
COACHING AND COUNSELING

The whole is greater than the sum of its parts.

—Aristotle

Synergy. The interaction of elements that when combined produce a total effect that is greater than the sum of the individual elements, or the sum of individual contributions.

—Dictionary.com

Synergy occurs when individuals work together in a positive and supportive working environment. When individuals work in positive and supportive environments, the organization reaps great benefits.

Every firefighter, fire officer, and chief stands on the shoulders of the legions of men and women who came before, who did the "JOB" and made it what it is today. Each generation of firefighters has left the profession better than it was and made it better through the service they gave (fig. 16–1).

My personal philosophy is that the "JOB" should always be, at the least, a little better when I got off duty than it was when I reported for duty. I have worked to ingrain this philosophy into the firefighters within my sphere of influence. This may be achieved by teaching, learning, fixing, cleaning, or repairing something; whatever you choose to do, it should make the "JOB" better. This is called making a contribution.

Fig. 16–1. The 9/11 Memorial in Weehawken, New Jersey, across the Hudson River from the World Trade Center site. We stand on the shoulders of giants. The mission is to honor them by leaving the "JOB" better than you found it.

The fire service is where synergy, cohesion, division of labor, specialization, dedication to mission, personal courage, and teamwork come together to serve, protect, educate, mitigate, rescue, and make safe the lives and property of the citizens we are sworn to serve (fig. 16–2).

A fire department is the equivalent of a Swiss Army knife for a community; that is to say, firefighters do it all and make do with whatever they have. We adapt, adjust, retool, repurpose, reinvent, and respond to every type of fire situation, every form of calamity, and the infinite number and variety of emergencies encountered in modern society. Anyone signing up for the job of firefighter is agreeing to be a productive contributing and dedicated member of an elite team of emergency operators. From the time a firefighter is sworn in to their first tour of duty and all the way to the end of their career, a firefighter is a member of a profession.

Fig. 16–2. The fire service is a career of learning, relearning, adjusting, and reevaluating. The fireground provides the challenge and often tests our dedication to our craft. (Courtesy Deputy Chief Mike Nasta)

Profession. *A calling requiring specialized knowledge and often long and intensive academic preparation; or the body of qualified persons in an occupation or field, or "THE JOB."*

Firefighters are required to follow, to the best of their ability, all lawful and service-related orders and direction. This is the fundamental contract every firefighter agrees to by swearing in good faith to serve and protect as a member of the fire service.

Fire officers understand their job. Officers are familiar with and adept in the technical knowledge, as well as the principles and practices used for confining, controlling, and mitigating fire, emergency, and rescue operations. Fire operators are at their best and most comfortable when they can see it, touch it, feel it, aim it, hook up to it, knock it down, cut through it, drive it, throw it, carry it, and so on.

It is often much harder to develop and apply the skills necessary to coaching and counseling. Nevertheless, a leader needs to have and hone coaching and counseling skills. Coaching and counseling comprise an essential part of a leader's workload. Enlisting the support and labor necessary to get any job done, meet any goal, and accomplish any mission is possible only when leaders invest time and energy in coaching and counseling the people placed in their charge.

Coaching and counseling skills do not come easily. These are sophisticated human interactions requiring maturity, compassion, empathy, personal courage, the ability to set boundaries, and a complete understanding of the roles and responsibilities of the participants. Coaching and counseling skills are acquired through study, practice, and then more study and more practice.

A fire department is a large team divided into smaller teams—platoons, divisions, battalions, and companies. Each of these component parts is placed under the command of an officer, chief officer, and chief of department.

In this book, we have been exploring management theories and skills that officers rely on to lead others. In addition to being a steward, a servant, a leader, a boss, a peer, and a subordinate, you, the fire service leader, are a coach and a counselor. Today's firefighter is a sophisticated, intricate, high-maintenance, complex, and sensitive asset—the most valuable asset to any department. The entire success of every fire, emergency, or rescue operation depends on synergy—individuals coming together, with each individual contributing and extending a combined effort to accomplish a common goal.

All fire departments require every facility, all equipment, and all apparatus to be inspected, inventoried, and serviced on a routine basis. The care and maintenance of this fire department "hardware" is codified within policies and procedures. Thus, care and maintenance of fire hardware becomes routine according to an established schedule.

Every time a firefighter reports for duty, PPE is situated and apparatus and equipment are inspected and made safe, in service, and ready. The company officer is responsible for overseeing this routine and ensuring that all hardware is ready to roll.

While a lot of care and attention is directed at ensuring the in-service and ready status of fire department hardware, it should be universally understood that even more care and attention must be given to the human assets—who are the "software" of the fire department. The in-service and ready status of the human component of a fire department must be a concern at the highest levels of authority. Responsible leaders must develop and implement conscientious methods to routinely monitor, tune up (train, educate, and support), and facilitate the in-service and ready status of the human components of their command (fig. 16–3).

Fig. 16–3. Subordinate readiness encompasses many activities. Coaching encompasses not only the hands-on activities seen here but also the mental, emotional, and motivational aspects of the "JOB." To this officer, this probationary firefighter must be the most important person on the "JOB." (Courtesy Alider Pratts)

Fire service leaders are given charge of educated, sophisticated, motivated, dedicated, and proud men and women. They must strive to create and maintain a work environment that is engaging, mentally and physically stimulating, purposeful, and fun. The leader must demonstrate commitment to the department, the mission, and the people they are charged to lead. None of this can be accomplished by simply waving a magic wand and proclaiming it leadership. Instead, a full-contact leader needs to mine the team for—and make use of—all individuals' contributions, whether they are talents, skills, personality traits, intellect, expertise, insight, or anything else that may be of benefit.

The tools available for the care and maintenance of fire department software are coaching and counseling. There are nine states of being in every fire and emergency department, each of which requires that fire service leaders coach _or_ counsel (or both):

1. Preparation
2. In service and ready
3. Standing by
4. In response
5. Arrival and size-up
6. In operation
7. Termination
8. Returning to quarters
9. Standing down

Coaching and counseling must be practiced and applied during every stage of fire service living and working.

Leader-Coach

There is no off-duty time when you're on duty.

—Flood

A leader-coach is someone who guides others to do, learn, and execute to the best of their abilities. In the fire service, leader-coaching is focused on developing the skills and knowledge necessary to safely and effectively perform the work required of a firefighter. A leader-coach instructs, reinforces, and clarifies the standards, requirements, and specifications of the job. A leader-coach strengthens and expands the contributive capacity of each individual team member to optimize the team as a whole.

Coaching requires having a plan. An effective leader-coach includes the team member(s) in the coaching plan and sets clear objectives. Coaching requires the leader to incorporate a feedback loop whereby team members can be listened to and acknowledged for accomplishment. Negative and positive assessment can thus be discussed and addressed.

Coaching is an exercise in interpersonal communication. The leader-coach must be hypervigilant as to *how* things are said. This is because how something is said very often is more important than *what* is being said.

Enthusiasm is contagious. Therefore, successful coaching requires the coach to be enthusiastic, as well as invested, engaged, encouraging, and supportive. In the fire service, good coaching encourages a supportive "Each one teach one" mind-set. Members of a well-coached team can be confident that someone always has their back. In the fire, emergency, and rescue business, good leader-coaching skills save lives (fig. 16–4). The success of leaders is measured by the success of the men and women they are charged to lead.

Fig. 16–4. The firehouse kitchen—there is no other kitchen like it. More training, educating, secret-hand-shaking, and team building happens in a firehouse kitchen than in any Ivy League school. (Courtesy Al Pratts)

Leader-Counselor

Counseling is interpersonal communication. This may take any of several forms, including (but not limited to) an interview, a job appraisal, a performance review, employee assistance, and support or guidance regarding a personal or job-related issue. Leader-counseling creates an interpersonal space where open and honest discussion of a subordinate's negative or distractive perceptions, behaviors, or actions can be identified and recognized as areas that need to be addressed. Counseling is used to clarify and reinforce team members' understanding of their roles in and responsibilities to the department. The goal of counseling is to align the needs, perceptions, behaviors, or actions of the firefighters with the needs and requirements of the formal organization. A counseling session is never focused on a person or personality; it is always focused on behavior and conflict resolution.

An effective leader-counselor must be able to separate personalities from the problems. A counseling session is a meeting between the team leader and a team member. The meeting is never about the person. Counseling must be confined to and focused on an issue, a pattern of behavior, a specific incident, or a particular aspect of an employee's performance. The focus of all counseling is to realign the needs of the employee with the goals and requirements of the department.

Leader-counselors should never confuse themselves with other specialists or advisors such as therapists, marriage counselors, physicians, priests, accountants, or fortune-tellers. Leader-counselors must understand that their function is as a reference to find other resources, by assisting employees to find and access problem-specific resources tailored to their needs. Thus, one of the most valuable statements to a leader-counselor is, "I'm not qualified to help you with that particular issue, but I can help you find someone who is an expert in that field. If you would allow me to help, I could check on that and get right back to you."

Resist the tendency to tell people what to do. A leader-counselor is a facilitator or advisor who helps subordinates recognize and identify specific problems. Once an issue has been defined, the leader-counselor can guide the firefighter through alternatives, options, and actions that may address or resolve the problem.

Under the best conditions, counseling should be understood to be positive discipline. A counseling session is often a proactive or preventive disciplinary action. In many cases, a counseling session is a form of intervention.

In an ideal world, every counseling session will result in a positive outcome; however, in the real world, it can go either way. A counseling session can go bad and result in negative disciplinary action when the employee refuses to take the suggested remedial actions or remains determined to continue the behaviors that initiated the session. When this occurs, the counseling session is documented and classified as an oral reprimand.

Counseling does not need to be confined to a formal office meeting. Leader-counseling is a skill that gives the officer a capacity to provide support and acknowledge employee efforts and positive contributions.

Counseling is a tricky business fraught with pitfalls. The up-close and personal demands of counseling have a discomforting effect on both the superior and the subordinate. The close living quarters and fraternal culture of the firehouse has the ability to muddy the lines between formal and informal. Interactions between leader and follower are often affected by personal relationships and friendship. It is easy for an officer to shy away from the requirements of being a representative of the formal organization. Coaching and counseling are the heavy lifting of full contact leadership. For most officers, coaching skills are easier to get their mind around, and counseling skills are more difficult to embrace and practice.

To be aware is to be alive.

—Plato

Counseling begins with supervision. The key to leader-counseling is being present, not only physically but also intellectually and emotionally. Successful counseling demands hypervigilance regarding the who, what, when, where, why, and how of the dynamics at play within the group. Effective counseling is the result of a leader doing much more listening than talking. An effective leader-counselor will do more asking than telling.

Counseling skills are often paired with coaching skills in three main forms: formal, semiformal, and informal (the last two of which are essentially the "lite" forms). Although some situations will benefit from on-the-spot attention, such *spot-counseling* should never be used to deal with an issue that requires a structured, formal, and confidential meeting between leader and subordinate. An effective leader will learn to recognize opportunities that require formal versus on-the-spot attention.

Just as fire department hardware requires regular tune-ups, software tune-ups are also beneficial. These are also perfect venues for a leader to make use of semiformal or informal counseling and coaching.

The same basic rules apply to all three forms of coaching and counseling:

- Keep confidentiality regarding delicate issues.
- Identify the problem.
- Gather the facts.
- Gain an understanding with the actor.
- Explain the consequences.
- Solicit feedback and input regarding the solution.
- Offer an alternative solution if necessary.
- Ensure understanding and reach an agreement regarding expectations.
- Document and follow up.

The counseling process can be initiated and executed at the company, battalion, division, and chief of department levels. Counseling is not a negative form of discipline. Rather, it is face-to-face communication between the officer or supervisor and the firefighter or fire officer. Leader-counseling should be understood to be a constructive forum. The focus for the leader is to provide feedback to the employee to correct the problem. All formal counseling sessions should be conducted in private.

Counseling skills and methods cannot be snatched from the air and successfully applied ad hoc. One-on-one, face-to-face interpersonal interaction is an art and a skill that requires tact, compassion, diplomacy, and a full and complete understanding of the roles and responsibilities of all participants and levels of command.

Subordinate counseling has five distinct components:

1. Recognize the importance of establishing an appropriate environment.
   - Use nonverbal communication to set the stage. Arrange furniture in a formal or less formal manner. For example, sitting behind a desk sends a completely different message than a more informal seating arrangement.
   - Greet the employee cordially and establish a rapport by asking a question regarding something other than the issue at hand. For example: "Good morning, Ben. Have a seat. How's that boat of yours?"
   - Identify the reason and purpose of the meeting.
   - If the employee is upset, angry, or frightened, a counselor will display sensitivity and understanding to calm the employee and defuse the situation.
   - Ensure that the employee understands that the meeting is the about the behavior, not the person.
   - Let the firefighter know that he or she is a valued member of the team and the department.
2. Understand supervisory responsibilities and demonstrate problem-solving skills.
   - Clearly identify the standards, expectations, and responsibilities required by the department.
   - Explain the need for improvement. Offer a path for achieving improvement.
   - Have the employee acknowledge that there was a problem. Then have the employee clearly state what the problem was.
   - Guide the employee in exploring alternative actions or methods for resolving the issue.
3. Maintain a command presence, composure, and self-control.
   - Express points of view or perceptions with clarity, persuasiveness, and poise.
   - Encourage open and honest discussion.
   - Demonstrate attentive listening. Examples of this are repeating what the employee has said and nodding to express agreement.
   - Pursue solutions to completion.

4. Have and demonstrate an understanding of the requirements of leadership.
   – An action plan for improvement should be agreed to in writing. The plan should include and specify any supervisory help or extradepartmental assistance.
   – A schedule and a set of benchmarks should be agreed on and included in the written plan.
   – A follow-up meeting should be scheduled to review progress or adjust action plan.
   – When all issues have been addressed and an action plan has been agreed on, the leader-counselor should restate and affirm the firefighter's value to the team and the department.
5. Perform self-assessment
   – Does the leader-counselor understand the consequences and implications of the decisions made and actions agreed on?
   – Are there other parties who need to be informed regarding the actions taken and decisions made?
   – Was the leader-counselor making firm and logical decisions?
   – Did the counseling session meet the requirements of the department's progressive discipline procedures?
   – Did the leader-counselor carefully document what transpired in the meeting?
   – Does a report need to be submitted to the leader's supervisor?

The following are basic guidelines for a counseling session:

- Do not put it off. As soon as all available information and facts are gathered, schedule the counseling session. Be prepared, have the facts, and carve out enough time to conduct an unhurried and thorough meeting.
- Focus on the issue(s). Be specific. Identify behavior and actions that are observable and measurable. Identify and clearly describe any unacceptable conduct.
- The member must be made aware of how the issues, actions, or behaviors affect the team and team members.
- Provide an opportunity for the subordinate to give his or her side of the story. Ask how he or she perceives what the issue is and

what the cause(s) may be. Ask what the employee thinks can be done to address the issue.
- Encourage the employee to speak freely and candidly. Listen carefully to the information given.
- The employee must be informed regarding what is acceptable work, behavior, or action.
- Question the employee as to what he or she thinks might be potential acceptable solutions.
- Keep an open mind. Consider all options.
- In the best-case scenario, the employee will offer up an acceptable resolution.
- If you think it is necessary, tell the employee what you think should be done.
- Show the employee how improving his or her behavior will be of value to the team.
- Give the employee a reason to improve work attitude. Offer suggestions—for example, an employee assistance program (EAP) or anger management counseling—to help the employee improve or change conduct.
- Reach an understanding about a corrective path of action.
- Clearly define your expectations for the employee's future comportment.
- Inform the subordinate that you will follow up. Describe the follow-up process and set dates including a firm date for the initial follow-up meeting.
- Meet again with the employee to review performance. Recognize improvements that have occurred.
- Take appropriate disciplinary action if the issue, behavior, or actions do not align themselves with the understandings agreed on during the counseling session.
- Inform the employee that the counseling session will be documented and confidential.
- Provide the employee with a copy of the report being filed.

Leader-counseling and leader-coaching should not be reserved for repair of the fire department software. Leader-counseling and leader-coaching are portable skills. Moreover, they are forms of mentoring.

Case Study: Something Real That Might Have Happened

Twelve-year veteran Firefighter Stew Dent, regarded as an all-round good and dependable firefighter, was the senior firefighter in the house. Cooking, cleaning, routine maintenance, and general all-around station duty were activities that Stew always found pride and satisfaction in. In an homage to the Beatles and Penny Lane, Dent kept his fire engine clean and well detailed. "He liked to keep his fire engine clean. It's a clean machine," he would warble at anyone within range. On the fire/emergency ground, Stew always pulled his weight and more.

After 12 years on the job, Firefighter Dent decided it was time to take a shot and study for the captain's promotional exam. This was to be Stew's first try. Like any serious studier, Firefighter Dent invested a good deal of time, money, and energy studying for the promotional exam. Post-test, Stew Dent was confident of success and happy to tell you that he had done well, certainly well enough to pass the demanding exam and be promoted.

When the promotional list was published, Stew's grade placed him in the lower 50th percentile. Even with his seniority, his chances for promotion were out of reach this time around. Within a week of the list's publication Stew's attitude and approach to the job began to change. He became sullen, withdrawn, and not as ready to lend a hand. Along with the attitude change, Dent got involved in a couple of out-of-character, lower-case conflicts with other firefighters in the fire station.

In-house scuttlebutt diagnosed and attributed Stew's sudden turn in attitude and behavior to disappointment around his promotional test results. The firefighters and the company officer didn't know what to do or what they could do, or they were just reticent to get involved. Leadership-driven coaching and/or counseling (formal or informal) were never insinuated into the Stew Dent situation.

The next shoe dropped at a multicompany drill. Firefighter Dent wasn't wearing his gloves. His company officer, Captain Fine, asked Stew to put his gloves on. Dent responded, "What difference does it make, not wearing my gloves? I'm only footing the ladder, I'll be all right."

Captain Fine reconfigured his "ask" into a clear and unambiguous direct order. Firefighter Dent countered by pointing out that Captain Fine was "being absurd." Captain Fine made abundantly clear his absolute disagreement with Stew's unflattering characterization of the order to don his gloves. The captain introduced a series of well-crafted remarks that crystallized for Firefighter Dent that his day had no chance of getting any better if he didn't jump-to and carry out the order to enclose his fire paws in a pair of NFPA-approved fire paw protectors.

All of this back and forth took place in front of other firefighters and officers. Reports were written and pushed up the chain of command, read, noted, and forwarded to the office of the division commander.

This Stew-stuff resulted in division-level discipline. Division Chief McDonald had a lot on her plate that particular morning but decided to squeeze the disciplinary meeting in instead of scheduling it when her work load thinned out. Division Chief Joyce McDonald met with the battalion chief and Company Commander Fine. The division chief listened to her officers report on the Firefighter Dent situation. She made a snap decision that Firefighter Stew Dent was to be transferred to another engine company.

What happened here could be characterized as the organization taking a cat-nap. In this instance the organization (represented by Deputy Chief J. McDonald) could be seen to have given up on Firefighter Stew Dent. The organization also failed at the company, battalion and division levels.

The beginning of a comeback

After the battalion and company officer were dismissed, Deputy Chief McDonald started to review and re-evaluate the Firefighter Dent situation as well as her decision to transfer the firefighter. McDonald knew that a transfer was not going to resolve the issue; it was only going to relocate it. In order to effectively deal with and get at the root of Stew Dent's behavior, someone needed to go to work in the uncomfortable face-to-face confrontation zone. Someone had to step up and "confront the uncomfortable."

When leaders are not trained or provided with the appropriate leadership tools and organizational support, officers will rarely be inclined to engage in the emotionally messy and uncomfortable moments that make up much of the day-to-day continuum of leadership.

Human-to-human interaction is complex, sophisticated stuff. Lack of focused, sophisticated human-to-human relationship training leaves leaders (and the organization) defenseless. Leaders need pointed schooling on how humans relate to and react with each other as individuals and as members of a group or team. Without a full portfolio of human relation–based leadership skills, leaders, supervisors, officers, chief officers, and chiefs of department end up trying to lead from behind a big fat curve.

The division chief realized she could have made the situation work in her favor and for the betterment of her command. She failed to turn the situation into a teaching/learning experience and a win-win instead of a loss-loss. Instead of guiding the company officer toward coaching and counseling, the division chief chose a nonaction-action, transferring a firefighter with an unresolved problem to another company.

Firefighter Dent's behavior was acknowledged, casually diagnosed, tolerated, ignored, and finally allowed to evolve into a free-range, overflowing problem. The Dent issue was flippantly decided upon and destined to be foisted onto another engine company that was unsuspecting, unprepared, unrelated, and undeserving of a "problem the cat dragged in."

Transfer as punishment often fails to calculate the disruptive impact that a "problem the cat dragged in" can have on the firefighters and officers stationed in the firehouse that the cat drags its problem into. Thus, Captain Fine would go back to his company-level leadership responsibilities none the wiser and even less equipped to handle the next uprising.

The comeback: righting the wrong

Deputy Chief Joyce McDonald didn't get to be the boss of a division's worth of firefighters, fire stations, fire apparatus, and firefighting equipment because she had collected the requisite clumps of chief dust. McDonald was one of the sharpest tacks in the firebox. The division chief was usually on top of her game and full-in on what was going on with her division and within the department. Today was an off day. Well, at least the morning was an off morning.

Deputy Chief McDonald reviewed what had transpired during the meeting. Joyce McDonald was not comfortable with how she handled the situation. McDonald gave herself a failing grade regarding her decision and disposition of the issue. During her personal after-action review, McDonald found herself guilty of lazy leadership. She blew it and she knew it.

And then she smiled.

(Note: *Lazy leadership occurs when a leader rationalizes away or fails to calculate the exponential negative consequences of half-stepping routine and/or seemingly incipient-sized exercises of managerial muscle or prerogative. "Lazy leadership" is a subject worthy of much scrutiny, including its significance, manifestations, and potential for playing havoc within an organization.*)

Division Chief McDonald saw the rarely recognized and the often unappreciated beauty of the "blew it, knew it" aspect of her personal post–bossing people around assessment episode. The beauty of "blew it, knew it" is in the gifting of a leader's mind with the permission and the capacity to change, readdress, teach, learn, start anew, stand up, stand down, give respect, get respect, and grow a command. In other words: the gift of full-contact leadership.

> *Leadership is a series of comebacks interrupted by a competing cavalcade of minor and major problems, tragedies, accidents, incidents, issues, situations, competing interests, and every other possible amalgam of humanly off-gassed flotsam and detritus capable of being ejected from the ever-oscillating fan of life."*
>
> —E. Flood

And Leadershipville is exactly where Deputy Chief Joyce McDonald went. The chief decided she would change her mind, once again proving that chief officers are equipped with minds designed for change when change is called for.

The first stop on the road to great leadership often starts at a pad of yellow, legal-length paper getting roughed up by a dull pencil outfitted with an industrial-strength eraser. With pencil and paper in hand and a high-quality eraser at the ready, Division Chief Joyce McDonald began to turn bad to good and wrong to right. In other words, she began to do bit of full-contact leading.

The first thing to do was reset her thinking. Joyce had to reclassify the entire situation and embrace it for what it truly was: a teaching, learning, and growth opportunity. The second thing to do was rescind the order of transfer. All battalions and companies would be briefed regarding the division chief's revised decision and the alternative actions being taken.

(Note: *Officers do not need to explain every action taken or order given, but every officer taking action or giving orders must be able to provide a rational justification for the action or order.*)

The division chief recognized that an early ministration of appropriate doses of leader intervention could have mitigated Firefighter Dent's issues before they turned into the series of problems that gave life to this case study. However, nobody in a position of authority was equipped to save Dent from himself. Joyce McDonald held to a pseudocalculation that estimated 80% of leadership is dedicated to protecting and untangling fire operators from the consequences of self-inflicted administrative grief. Division Chief McDonald determined that her company and battalion commander's coaching and counseling skills needed to be assessed, enhanced, and then supported.

Expanding the sphere of influence

Division Chief McDonald determined that a crash course focused on the leadership skills required to conduct coaching and counseling would be developed and delivered to units at the battalion and company level of her division. Each battalion chief would be given a reasonable but abbreviated timeline to create and submit for division approval, lessons plans, and training materials for conducting assigned classes as directed. Battalion 1 would be responsible for a program on *Maslow's Hierarchy of Needs*—understanding the motivation behind a person's behavior. Battalion 2 would be responsible for a program on *group dynamics*—understanding how people act and react within groups. Battalion 3 would be responsible for a program on *transactional analysis*—how to develop successful and effective communication skills. Division Chief McDonald would be responsible for tying the whole program together. Her task was to develop a program that could demonstrate how Maslow, group dynamics, and transactional analysis could be introduced into a real-world situation going bad.

Properly understood and surgically wielded, leadership skills are the safest, most effective path of least resistance to confronting uncomfortable situations and wrongheaded behaviors *before* the situations or behaviors can metastasize into problems of larger scope.

Deputy Chief Joyce McDonald was under no illusion that a "crash course" would be the end-all and be-all of the issue(s). She would use the work done by the officers and firefighters of her division as a template and a first step toward enlightened maintenance of the human-type fire equipment. She would run (sell) her program idea by the chief of department and chief of the training division to get the necessary juice to coordinate with and tap the resources of the training division and any administrative entities deemed necessary.

If McDonald's "in-house" program proved successful, she would provide the chief of department and the division of training with a full formal report, lesson plans, and the recommendation that a program focused on coaching and counseling skills would be of benefit department-wide.

Some other thoughts on the subject

Firefighters and fire officers are trained at a high level of proficiency to troubleshoot and perform preventive maintenance on a wide variety of tools, equipment, and apparatus. To accommodate this, departments issue equipment maintenance schedules and operator specifications. Further, levels of supervision are in place to monitor and ensure the in-service and ready status of equipment and tools. Routine preventive tool and equipment maintenance exists because it is cost-effective, really smart, and the results are positive and measurable.

When a firefighter finds an SCBA cylinder low on air, the cylinder will be pulled, charged, and returned to service. Firefighters are not only trained to troubleshoot a problem, they are trained to take specific actions to ensure in-service ready status. No right-minded firefighter would ever consider to be an acceptable solution the transferring of the low-air cylinder to another fire house.

The scenario presented a well-considered veteran firefighter whose expectations had not been rewarded with great success. Anyone who's gone through the test gristmill understands, and anybody who's been in proximity of a test taker can empathize as well.

Firefighter Dent was withdrawing, argumentative, and acting out of character. His behavior crescendoed at the drill site. Dent's behavior finally got him what he wanted or needed. What he wanted or needed was attention (not the kind he got). He didn't know how to ask for help, and his superiors were unable to troubleshoot and then deal with the situation. If Firefighter Stew Dent was an expensive piece of equipment presenting imminent malfunction, he would have been assessed and action would have been taken to make certain that his in-service status would be certified.

Leaders need be trained and retrained (and then trained again) in the routine maintenance and proactive supervisory practices (supervision) required to troubleshoot, recognize, and address problems before they become the purview of a higher authority. Fixing a problem where and when it rears its head creates a perfect venue for on-the-spot teaching and learning opportunities. Effective, engaged leaders are opportunists and mentors always looking to turn a problem into an educational opportunity.

In simple short form, routine preventive human maintenance looks something like:

Step 1: Monitor (your people).

Step 2: Assess (in-service ready status of your people).

Step 3: Make ready and certify in-service ready status (of your people).

It is a matter of fact that fire women and men are not only the most valuable asset in an organization; they are also the most expensive asset and represent the department's best investment. An enormous volume of time, attention, treasure, and human effort are dedicated to ensuring the service dependability of tools, equipment, apparatus, and facilities, which is as it should be. Serious and committed investment in human maintenance practices need be commensurate with the fire service's investment in its leaders. If success is a goal, it is imperative for a fire department to support and maintain an engaged, effective, and mission-centric leadership corp. The return on a real trained-to-the-bone type investment in a career-long officer/leader support and development program could be staggering.

Transfer defined

Transfer is a valuable managerial tool, a requirement in every leadership skill portfolio. Transfer is necessary to the effective and efficient operation of any fire/rescue/emergency organization. Transfers are intended to support and facilitate staffing and coverage requirements. Transfers provide the organization the necessary flexibility required to meet staffing and coverage needs.

Temporary duty assignments and relocation of unit(s) are forms of transfer. Promotion or change of title can be cause for transfer. A member, by way of formal request for change of assignment, can likewise initiate a transfer, promotion, or change of title.

Transfer is also a staffing tool. Transfers (like all management tools) are called upon to improve, support, and protect the organization, its membership, and its mission. Using a staffing tool to resolve a behavioral issue is as sophisticated as telling someone to "Go stand in the corner!"

Sometimes a mix of personalities, an unevenness of skill and knowledge levels, personal issues, and other day-to -day complications can make transfer decisions necessary and appropriate. At times, common normal life factors and external forces may find appropriate resolution in a transfer. These types of transfers happen, are acceptable, and are not considered in the same category as transfer as punishment.

Transfer is change. Change creates stress. Change should not be introduced casually. Managing change requires a real need for change, forethought, empathy, and diplomacy. Most importantly, a rationale for the change must be communicated. A leader does not have to justify every order. However, a leader must be able to provide a sound, reasonable justification for any order given or action taken. (Yes, you're correct; this has been stated a couple of times already. How come? You may ask. Because it's important! That's how come.)

Discipline vs. transfer as punishment

The prime directive of leadership skills/tools is: A leader's actions and orders must always serve the purpose of supporting and protecting the member, the organization, and advancing the mission of the fire service. The word discipline comes from the Latin *disciplina*, which means instruction given, teaching, learning, and knowledge.

Discipline is defined as training that corrects, molds, or perfects behaviors and mental faculties. Discipline is the practice of training people to obey rules or a code of behavior, using education and progressive forms of censure to correct and/or redirect errant behavior.

The best discipline policies are front-loaded with positive and supportive education, training, guidelines, and clear unambiguous prescriptions for action. Sanctions should be the option of choice only after the organization and its representatives (officer corps) have explored all available positive discipline alternatives.

Punishment is defined as the infliction or imposition of a penalty as retribution for an offense. Punishment and discipline are not synonymous. Punishment is censure and retribution.

Negative disciplinary action is a censure, an expression of disapproval or a harsh rebuke, designed to teach or realign an employee's behavior with the goals and responsibilities of the department, not to seek retribution.

All negative forms of discipline should be designed to re-educate the member(s) and dove-tail the needs of the member(s) with the goals and objectives of the company, battalion, division, and department. All negative disciplinary action should flow from an informed and thoughtful decision chain. Negative discipline must never be casually meted out or designed to be vengeful.

Negative discipline is the last card to be played by a supervisor. Positive and negative disciplinary actions are a set of personnel protective devices. Positive discipline is demonstrably more effective and much more cost-effective than negative discipline. Negative discipline should be introduced *only after* the organization has exhausted all the positive front-end alternatives and remedies.

Fun facts regarding transfer as punishment

Transfer as punishment is managerial sleight-of-hand. Transfer as punishment only pretends to deal with an issue. This pretense allows leadership to feel like the situation has been dealt with (pretense leadership?). Feeling like a leader is not the same as actually being a leader. That is not a feeling, that is a fact.

Transfer as punishment is a failure of leadership, a white flag of surrender. Punishment transfers are often knee-jerk, reactionary, disruptive, and almost always ineffectual.

Discipline is a team-building tool. Transfer as punishment is not a team-building exercise. Transfer as punishment is a form of shunning that runs contrary to the concepts of unit cohesion and team building. Transfer intended as a punishment is equivalent to moving the seat of a fire to the basement so you don't have to look at or deal with it. And while you're not looking you are fervently hoping the seat of the fire will just go away

Transfer as punishment is a disruption. Disruption ripples exponentially, in bad ways, up, down, and sideways through the chain of command. As can be seen, a lot can be said, but not much good can be said about transfer as punishment.

chapter 17

CASE STUDY: THE SENIOR GUY

Captain Tim Finnegan is the new company commander of Engine 203. He was promoted four months ago and recently assigned to one of the busiest houses in the department.

There are three firefighters assigned to Engine 203:

- Firefighter Andrew Slip is a 24-year veteran who has worked at Engine 203 for 20 of those years. All of Slip's past evaluations have been excellent, but word is that Andrew may be retiring as soon as he gets his 25 in. In his off time, he enjoys boating and is an avid fisherman. He also has twin boys who will soon be graduating from high school and are applying to colleges around the state.
- Firefighter Bill Lemonie has a little more than two years on the job. Lemonie's evaluations have been excellent, and he is one of those people who is enthusiastic about training and always glad to help out. He is an all-around nice guy.
- Firefighter Deborah Harry is a probationary firefighter just four months out of the academy. Deborah finished in the top 10% of her class at the academy.

The previous captain made it company policy to assign the senior firefighter to the position of driver and pump operator. As the senior firefighter, Andrew Slip was assigned this role under the previous engine officer and has held the position for the past five years.

Captain Finnegan had a different philosophy from the previous engine officer. Finnegan wanted the senior firefighter to work closely with the probationary firefighter, buddying up at all incidents and providing her with as much help as he could. Captain Finnegan also believed that every firefighter should be fully qualified and experienced at every engine company position. Captain Finnegan explained to the members of his crew that there was no such thing as a specialist to him; every firefighter would be required to cover every position and do every job required of the engine company.

Finnegan reviewed Firefighter Lemonie's training records. He assessed Bill's driving skills when on squad calls and when the company was on assignments other than emergency response. The captain also evaluated Lemonie's ability to "get water." Lemonie passed all evaluations with flying colors.

Captain Finnegan kept the firefighters informed about the what and why of the changes he intended to implement. He told the firefighters that he would call a company meeting prior to implementing new assignments.

As soon as Captain Finnegan was comfortable with the capability of the members of his company, he held a company meeting, during which he outlined his plan and the rationale behind it. He informed all members that the riding assignment changes would take effect on the next tour of duty.

Firefighter Lemonie and Probationary Firefighter Harry were enthusiastic when the assignments were made official, but Firefighter Slip, not so much. Throughout the meeting, Slip was rolling his eyes, shaking his head, and smirking.

During the meeting Firefighter Slip pointed out that he had seniority and that his previous boss gave him the driver and pump operator job. Slip also believed that the "senior guy thing" was company and department policy—and if not policy, certainly a tradition. Andrew's tone of voice was argumentative and angry.

Captain Finnegan responded by telling Andrew that he understood where he was coming from and that he was willing to discuss the issue. However, they would discuss it in private in the captain's office.

"Forget it!" Andrew slapped his palm onto the kitchen table and stomped out of the room.

Unflustered, Finnegan asked the remaining firefighters if they had any more questions regarding their new assignments. Neither Bill or

Deborah had any questions. Finnegan dismissed the two firefighters and walked upstairs to his office.

Captain Finnegan understood that no good would come of confronting Firefighter Slip at this particular moment. The captain would allow Slip some time to calm down and hopefully assess his behavior at the meeting. Even so, Finnegan also recognized that the situation could not be allowed to stand and must be addressed as soon as practical, that very day.

Captain Finnegan took good advantage of Firefighter Slip's cooling-down period to create a framework for the leader-counseling session he would hold with Slip (fig. 17–1). Once Captain Finnegan had completed his outline and gathered his thoughts, he called Firefighter Slip up to his office.

How did the interview go? As you can see from the following play-by-play recounting of this counseling session, it went well. To make it easier to follow the conversation, Captain Finnegan's dialogue will be presented in boldface and Firefighter Andrew Slip's responses in italics.

Come in, Andrew. Take a seat. How are those twin boys of yours?

They're good, Cap. Both boys are doing real well in school and starting to look at colleges. Thanks for asking.

I'm glad to hear they're doing so well. Kids grow up fast. You must be very proud of your boys.

Yes, Cap, real proud.

Do you have your boat in the water yet?

Not yet. It's still a bit too chilly. Couple more weeks and I'll put her in.

Something you're looking forward to, I'm sure.

Yes. Actually I can't wait.

Andrew, the reason I've asked you to my office is to discuss my concerns regarding your behavior during and at the end of the company meeting this morning. But, before we get into that I want you to know that

I have reviewed your record. You're a veteran firefighter. Your record is sterling and all your evaluations rate you as an excellent firefighter.

Andrew, by all accounts, you are an excellent firefighter, a senior guy with lots of experience. I want to tell you that you are an important and valuable member of this company. You can be the kind of firefighter that the younger firefighters are able to look to for guidance. You're a key member of this company. You have a lot to be proud of and a lot to contribute.

> *Thanks, Cap. That's very kind of you to say.*

Again, we are here to discuss the outburst and stomping out of the meeting before being dismissed. Could you explain to me what was going on with you during the company meeting? Can you tell me what precipitated the angry outburst, slamming your fist down on the table, and the abrupt exit?

> *Like you said, Cap, I'm the senior guy. I'm used to doing things the old way. I liked the way things used to be. Even though we got a new boss, I just figured that I would keep my job, driving and pumping, but you changed all that. Everything has changed in this firehouse—new boss, new firefighters, new assignments, new everything.*
>
> *I don't know, Cap, things are different for me nowadays. The job just ain't the same. These new kids are a different breed. Sometimes I just don't get it. Maybe I just don't fit in anymore.*

Andrew, I want you to know that I understand what you are telling me and how you are feeling. The fire department, just like the rest of the world, is constantly changing. Me, you, and everybody else has to deal with that reality. As Battalion Chief Browne says, "There is nothing more permanent than change."

I want you to know, to my mind, everybody in this company fits in. You fit in, Deborah fits in, Bill fits in, and yes, even me, I fit in.

Let me ask you: Do you believe that the outburst and your rushing out of the kitchen was the best way to address or deal with the issues you've just brought up?

> *No, boss. Not really. Of course that is not the way I want to do things. I just lost it. I apologize.*

I accept your apology, Andrew. However, we need to properly address certain things. First I'm going to ask: Do you recognize that slamming the table and busting out of the kitchen are violent, hostile, disturbing, and disrespectful behaviors?

> *I really didn't think about it in that way. I'm a hot-blooded guy. Sometimes I lose it. I figure, "It's the firehouse. They'll get over it."*

Andrew, I totally understand. And I believe everyone has the right to get upset and disagree. However, no one has the right to disrupt the workplace or act out in a violent or hostile manner. That kind of behavior is disruptive and causes stress and anxiety among the members of the company. It is disrespectful to the team as a whole and the officer in charge.

Can you see my point? Do you understand what I'm saying? Can you put yourself in the shoes of your fellow team members?

> *I hear what you're saying, boss. I really do.*

How would you feel if you were discussing something with your family and one of your boys responded in an angry and dismissive manner, banged on your kitchen table, and then just walked out on you, your wife, and your other son?

> *I wouldn't like it a bit. I don't stand for that kind of behavior in my house.*

I need you to tell me that you understand your behavior was unacceptable and disruptive to company unity.

> *Yes sir. I clearly understand that my angry outburst was not acceptable behavior and I understand that it had harmful effect on the company and the firefighters who witnessed my outburst.*

Thank you, Andrew. So, it seems, you can understand how the rest of the company might feel regarding the behavior that occurred at today's meeting.

> *You got me there, Cap. I can see how inappropriate, disruptive, and immature my actions were.*

Well, Andrew, I did not get you there. You came to the realization on your own. Andrew, you also said, "They'll get over it." Do you really believe that the firefighters you work with should just get over it?

> *Actually, Cap, I think I owe them an apology and an explanation. I'll get on that as soon as I go downstairs.*

I absolutely agree with you, Andrew. Let me ask you the following: As the wise veteran of our company, what additional action would you advise someone in a situation similar to yours?

> *I don't know if I have much advice, but I do have a story about my days as the new kid on the block.*

Fair enough. Tell me what it was like for you when you were new.

> *When I first got on the job and was the new kid, a guy with a lot of years and experience took me under his wing. He showed me the ropes, kept an eye on me, and really helped me out. He was a great role model, and I was a better firefighter for his efforts and attention. I was always thanking him, but he was not a touchy-feely kind of guy. He would always tell me the same thing: "Talk is cheap, kid. Action speaks louder than words. Veteran guys have a responsibility to support, educate, mother-hen, and when necessary, wipe the noses of young ones." He also told me that whenever it became necessary, senior firefighters would enlighten every Johnny-come-lately regarding the cold hard facts of life in the fire department. The thing that he always preached was this: "Remember, kid, when you go home in the morning, the "JOB" should, in one way or another, always be better off 'cause you were there the day before." He never stopped reminding me that I was only holding a place for the next new kid coming up. And this discussion is making it clear to me that I've been holding a place for the two new kids downstairs. He always told me, "Make it your mission to learn something new and useful every day. Make an effort to help someone else learn something new and useful every day. Do the right thing and you will cast a long shadow on this "JOB"."*

Well, Andrew, you were very fortunate to have worked with a guy like that. Are you familiar with the saying "paying it forward"? Because that is exactly what your story was all about, paying it forward.

> I certainly was fortunate. That guy made a big difference in how I came up in the job. I seem to have forgotten how important it was to for me to have someone wipe my nose and kick my butt when I needed it. I needed all that and a lot of other stuff. Anyway, I get it, Cap. I've got to give back. I want to give back.

Andrew, you have plenty of time and a great opportunity to pay it forward and to do what was done for you. Let's summarize, talk about moving forward, and define where we are and what we're going to do. First, you've told me that you understand that hostile and disruptive behavior is unacceptable and inappropriate. Is that correct?

> Yes sir, I do understand.

You further agree that the behaviors discussed here today are unacceptable and harmful to the team and its members. If this behavior is repeated, it is clear to you that disciplinary action is the next step?

> Yes, I do, sir.

You have agreed that Bill and Deborah are owed an apology and an explanation. You agree that you will take care of that as soon as this meeting is concluded, correct?

> Yes sir.

Now I want to address what I consider the most important issue. You described an amazing and extremely positive relationship between a solid, experienced, and right-minded veteran and a newly minted firefighter. My question is this: Do you think that you could do for Bill and Deborah what was so generously done for you?

> Definitely sir! I'm on it!

So we both agree that you have a great deal to bring to the table. And we have an agreement that you will commit to mentoring the two younger members of Engine 203?

> *Absolutely, boss. I see it as a responsibility and an opportunity to give back what was so generously given to me. I won't let you down.*

I want to speak to the assigning of riding positions. You understand that any and all decisions regarding the assignment of riding positions falls under the authority and is the responsibility and prerogative of the company commander.

> *I do, sir.*

One more thing: I want you to know that I am willing to meet with and am available to all members of this company to respectfully discuss, review, consider, and explain any issues or concerns regarding this engine company and its operation.

> *Thank you, sir.*

Very good, Andrew. I appreciate your frank and honest discussion of the issues at hand. I am impressed with your willingness, and I expect your efforts and contributions to have a positive effect on this company and the junior members of Engine 203.

I want to inform you that this meeting is considered a formal counseling session. I've made notes and will document this meeting. The report will include what has been discussed, what we've agreed upon, and the expectations that have been defined. I will provide you with a copy of the report that I will be filing. What has transpired in this meeting will be kept in strict confidence. We will have a follow-up meeting in one week. At the follow-up meeting, we will review your progress and address any issues you may have.

Andrew, I have an open-door policy. I will always make the time to speak to any member of this company regarding any issues or concerns they may have. Before I end this meeting, I want to once again thank you for your cooperation and reinforce the fact that I have great confidence that you will hold to all that has been agreed to here today. Your contribution as senior firefighter will have an important and very positive impact on the junior members and the company as a whole. This meeting is concluded. You may return to your duties. Thank you for your time and attention.

> *Yes sir. Thank you, Captain Finnegan.*

chapter 17 | Case Study: The Senior Guy

A well-conducted coaching and counseling session is a sophisticated interpersonal form of communication. A lot transpired during the coaching and counseling session between Captain Finnegan and Firefighter Slip. The following questions are intended to guide the reader in analyzing the transcript from this hypothetical coaching and counseling meeting. These questions are designed to highlight many of the transactional subtleties found in the above exchange:

- What styles of leadership were in play?
- What leadership skills did Captain Finnegan use to defuse any resentful or potentially combative behavior that may have found its way into the meeting?
- Were the principles of transactional analysis effectively used?
- Regarding Maslow's hierarchy of needs, what human needs were addressed and met?
- Can different types of power be recognized?
- Was command exercised?
- How would Captain Finnegan's command presence be characterized?
- Was the meeting well structured?
- Did the meeting have a well-defined beginning, middle, and end?
- Was the tone and tenor of the meeting conducive to reconciling and resolving the issues explored?
- Was there more asking than telling?
- Did Captain Finnegan's counseling skills facilitate and guide Firefighter Slip to self-assess?
- Did Captain Finnegan allow Firefighter Slip to recognize the impropriety of and the negative impact his behavior had on company unity?
- Did Captain Finnegan ensure that Firefighter Slip was aware of his behavior?
- How did Captain Finnegan ensure that Firefighter Slip understood the consequences of his behavior?
- Did Captain Finnegan's counseling and coaching skills tap into Firefighter Slip's personal experiences in a manner that allowed Slip to reconcile the realities of change?

- Was Firefighter Slip encouraged to find a way to resolve the issues?
- Did Captain Finnegan focus on behavior instead of personalities?
- Did Captain Finnegan acknowledge and support the positive potential that Firefighter Slip's seniority and experience could bring to the team?
- Was the meeting a team-building experience for both the officer and the firefighter?
- How helpful was it for Captain Finnegan to sit, think, and consciously create an outline for the counseling session?
- How would the outcome of the meeting be characterized?
- What goals were defined during the meeting, and were they properly addressed?
- Did the officer help align the needs of the firefighter with the needs of the department?
- How did Captain Finnegan fare as a representative of the formal organization?

The complexity of human interaction and its potential to have a profound impact are what makes it worth serious consideration. Leadership is humans interacting with other humans. What else is there?

Readers are encouraged to look to their firehouse experiences, create scenarios, and outline the counseling session they would conduct when confronted with situations that required confrontation between a leader and a follower. Role-playing and simulations are powerful methods for honing coaching and counseling skills.

Coaching and counseling are by definition confrontations. Conducting mature, goal-oriented, respectful, employee-centered confrontation is an important responsibility of the full-contact leader. An officer must exhibit a command presence; command confidence is based on command competence. A firm, fair, and friendly demeanor is the tact a leader takes when carrying out the responsibilities that come with leadership.

chapter 17 | Case Study: The Senior Guy 201

Subordinate Counseling Worksheet

Review Records: _____

Date: _____

Time: _____

Officer conducting the counseling: Name, Rank & Assignment: _____

Member being counseled: Name, Rank & Assignment _____

Other parties or witnesses: Names, Rank & Assignment _____

Set up the room.

Take a moment to review your interview strategy

Take appropriate action to eliminate disturbances/nonemergency calls, etc.

Clear desk of any clutter

Ensure all required documents, papers, and note-taking material are at hand

Place the firefighter's chair in front of the desk directly across from the interviewer

Opening:

Greet the member, shake hands, and direct the member to take a seat

Set the proper tone: Firm, fair & friendly

Small talk (general and/or background information)

Body of the Counseling Session:

Outline purpose/reason for meeting: _____

Discuss positives/experience, past performance & evaluations: _____

_____

Identify Issues/Develop a Plan of Action

Clearly and Honestly Describe the Exact Behaviors That Brought About the Meeting

Issue #1 _____ Plan _____

Issue #2 _____ Plan _____

Issue #3 _____ Plan _____

Gain agreement: _____

Close of Session:

Summarize & review: _____

Inform the member that the meeting will be documented: _____

Schedule follow-up meeting: _____

Inform the firefighter re: open-door policy: _____

End meeting & dismiss the firefighter: _____

Fig. 17–1. Captain Finnegan used this worksheet to create a framework for his leader-counseling session with Firefighter Slip.

chapter 18

CASE STUDY: THE POWERS PREINSPECTION—PART I

You Know It When You See It

The Grace Powers case studies are presented as a series of interrelated stories. The intent throughout this series is for the reader to be able to watch as the exponential factor drives the events that unfold.

These case studies begin in the office of Division Chief Strong and follow Battalion Chief Grace Powers through a number of situations including supervisory assignments and leadership activities. Through their example, the reader is encouraged to compare command and leadership approaches. Readers are invited to debate the advantages and disadvantages of each approach, as well as to identify methods or skills that they would want in their personal leadership portfolio.

Meeting with Division Chief Strong

In a meeting with Division Chief Strong, newly minted Battalion Chief Grace Powers confided that she felt tentative and uncomfortable while making routine, assigned rounds to the firehouses under her command. The battalion chief explained that she felt like an intruder and trespasser.

The division chief thanked Powers for bringing her concerns to his attention. Division Chief Strong recognized the value of a command-level officer who was capable of honest self-assessment. Being able to ask your boss for help requires self-confidence and courage. Deputy

Chief Strong told Powers that he appreciated her honesty and understood what it took to come forward with her concerns.

Leadership guidance

The deputy chief offered the following guidance to Powers: "A battalion commander is personally charged with the care and safekeeping of all facilities, apparatus, equipment, and personnel assigned to their command. Every building, all facilities, every apparatus, every piece of equipment, and the readiness of all personnel fall under the battalion commanders' authority and are their responsibility." The deputy chief went on, telling the new battalion chief that she, in essence, "owned" everything she commanded; every station, all apparatus, and every piece of equipment should be treated and recognized as her property and her responsibility. Other than personal lockers, there should be no closet, room, compartment, nook, or cranny to which she should not allow herself access.

A battalion commander will take credit when it is due. Conversely, a battalion chief will take complete responsibility when all the stuff comes into contact with the ever-oscillating and unforgiving blades of the "fan of command." A command officer needs to accept all of the responsibilities that come with a command-level position.

Ceding inappropriate power to junior officers or the informal group will dilute the authority of the formal leader. Timidity and indecision are hallmarks of diluted leadership. Diluted leadership is even more dangerous than absentee leadership.

Unambiguous boundaries must be drawn to define where the influence of the informal group ends and the authority of the formal group begins. The fire and rescue business is not baseball—and, thus, the tie does not go to the runner. When in doubt, the leader will represent for the formal organization.

The dynamics between formal and informal groups must be understood and harnessed to yield effective and productive results. A battalion chief is a team builder. Leaders need to cultivate a strong and confident command presence. To effectively lead, command-level officers must exploit all opportunities to align department needs with the needs of the firefighters and fire officers under their command.

Overview

The battalion chief had described a lack of confidence. Strong knew that a frontline commander cannot be insecure around routine assignment in softer environments. Insecurity in soft environments is almost guaranteed to carry over to harder environments. Indecision or insecurity at a command level will translate into confusion and, consequently, unsafe and dangerous conditions on the emergency ground.

Leadership development is always a priority. Battalion Chief Powers's situation provided a teaching, learning, and leadership development opportunity. This harkens back to the motto, "Each one, teach one." Deputy Chief Strong's actions would both develop the battalion chief's leadership and educate the captains. Deputy Chief Strong would craft an assignment allowing Battalion Chief Powers to project an enhanced command presence. This assignment would simultaneously demonstrate the support and the confidence Division Chief Strong had in Battalion Chief Powers (fig. 18–1).

Fig. 18–1. Battalion commanders must understand their roles and responsibilities and serve both those above and those below battalion rank. Their task is to funnel critical information from both directions to align the goals of the formal organization with the goals of those under their command. (Courtesy Ron Jeffers)

Leadership development is both a bottom-up and a top-down experience. Every human transaction is rich in lessons to be learned. Leaders must be able to learn from those they lead, as well as from those they follow. Effective leaders embrace the notion that they are lifelong students and, therefore, continually seek to hone the skills and better the practices necessary for good leadership. Creative and effective leaders are always open to learning from their peers, their subordinates, and their superiors. Because Battalion Chief Powers came to Deputy Chief Strong with her concerns, the deputy chief recognized that regularly scheduled officer meetings would be of great benefit to his entire division. Holding monthly meetings would provide developmental opportunities to enhance communication between leadership levels. Because Deputy Chief Strong appreciated that he was made aware of the need for regular officer meetings, he made it a point to thank Battalion Chief Powers for bringing this need to his attention.

The deputy chief decided that he would begin holding officer meetings on a regular monthly basis. A monthly officer's meeting is a team-building device. Such meetings create a forum to bring forward ideas, recommendations, issues, and solutions. Putting all heads together produces more ideas and better results than any single head alone.

Inspection Assignment

Deputy Chief Strong concluded his meeting with Battalion Chief Powers by informing her that one way he would address her concerns would be by having her conduct a formal and comprehensive inspection of all fire stations within her battalion. This assignment would require Battalion Chief Powers to project a strong and confident command presence. Powers would involve her company officers in the planning, organizing, and scheduling of the inspection. The success of the assignment demanded that Battalion Chief Powers engage with her officer corps through arranging, supervising and following up on all aspects of the assigned inspection.

Deputy Chief Strong understood that it was necessary to create a supportive framework for Powers's inspection assignment. Thus, Strong would provide complete and specific direction outlining the methods to be used, the goals to be achieved, and all expectations regarding the inspection. Specificity of direction would reduce or eliminate the potential for timidity and indecision.

Deputy Chief Strong instructed Battalion Chief Powers to prepare herself and the companies for the inspection. Battalion Chief Powers was ordered to conduct a preinspection planning meeting with all company officers. All companies were to be informed of and given adequate time to prepare for the inspection, and Battalion Chief Powers was to clearly state her expectations regarding this assignment through a verbal announcement to every company; she would then follow up with a written notice, which would include a schedule of inspection.

Deputy Chief Strong directed the battalion chief to determine the length of time necessary to complete an inspection in advance so that she could schedule her time accordingly. Preinspection preparation was to include a briefing for the division chief.

A standard station inspection checklist would be developed and presented for Deputy Chief Strong's approval. If the checklist was found to be acceptable, the division chief would forward a copy to the chief of department with a recommendation that the inspection checklist be adopted department-wide.

Deputy Chief Strong directed Battalion Chief Powers to have the assigned company officers accompany her during inspections. In the future, all company officers would be required to accompany the battalion chief during routine station rounds. Company officers were to supply the inspecting battalion chief with all inventories, repair reports, and all in-process supply and maintenance requests.

The inspection would focus attention on four categories: facilities, apparatus, equipment, and personnel. Each category was to be inspected individually.

The inspection created opportunities to acknowledge positive results. Conversely, the battalion chief would require the remedy of any unacceptable conditions found to be within the control of the company. Issues requiring support beyond the company level were to be documented. The battalion chief would take the necessary actions to address any such issues (fig. 18–2).

Fig. 18–2. Command presence—you get it by earning it. (Courtesy Becki White)

Battalion Chief Powers would conduct follow-up action(s) during future routine rounds. Once Powers had left his office, Deputy Chief Strong took pen to paper and wrote a synopsis of his meeting with her.

After-Action Analysis

Deputy Chief Strong's command philosophy and vision guided his actions, orders, and expectations regarding the issues brought forth by Battalion Chief Powers:

- Command and company officer development were the focus and desired outcome regarding the assignment given Battalion Chief Powers.
- Deputy Chief Strong demonstrated support, compassion, and understanding.

- The goal of Deputy Chief Strong's assignment was to ensure dependable, capable, confident, and effective command staff.
- Deputy Chief Strong provided a path for the new battalion chief to develop self-confidence, as well as to project a command presence.
- Battalion Chief Powers brought forward an important and difficult-to-disclose issue.
- Deputy Chief Strong recognized the potential negative consequences this issue presented.
- Deputy Chief Strong addressed Battalion Chief Power's confidence issue by assigning a success-oriented task. Within a supportive framework, Strong directed an assignment designed to empower his newest battalion chief.
- The inspection assignment would bring Battalion Chief Powers into a structured and purposeful interaction with the companies under her command.
- The assignment was clearly defined and success oriented. Setting a schedule, creating a checklist, and briefing the division chief provided benchmarks for success. Everyone from firefighter to company officer, battalion chief, and deputy chief was invested in the inspection.
- Every level was positioned to receive positive feedback.
- The inspection was to be performed in a structured, comprehensive, and professional manner. All parties concerned were provided with lead time, specifications, and a clear set of expectations.
- Battalion Chief Powers was provided with a detailed prescription for action designed to create a command experience that bolstered her confidence. The assignment would send a clear message that Powers was the boss.
- The process would encourage the battalion commander to give herself permission to access all areas of her assigned command with confidence.
- The entire experience was educational and productive.
- The officers and members of the company, battalion, division, and department were qualitatively and quantitatively better off as a result of their efforts and the experience overall.

chapter 19

CASE STUDY: THE POWERS PREINSPECTION—PART II

You Know It When You See It: Preinspection Preparation

As ordered, Battalion Chief Powers held a planning meeting with her company commanders. This meeting provided her with a formal and controlled environment in which she could interact closely with her company officers.

The division chief's inspection assignment was designed to allow Powers to build confidence and demonstrate a strong command presence. A formal structured preinspection meeting thus was a framework within which Powers could exercise command and control—a safe and supported context that would define and reinforce both her authority and responsibility within her command (fig. 19–1). Including the company officers in the inspection planning served as a team-building exercise and helped to create a vital group image within the company officer ranks.

During the meeting, it became evident that some preparation tasks would require the support of line and staff positions outside the battalion. However, the company officers noted that some requests for support could cause duplication of efforts. Battalion Chief Powers agreed with their assessment, and she responded by having her company commanders identify any areas where there would be duplication of efforts. Once these points were identified, she had each of the company officers choose a task assignment. Thus, each company officer was responsible for ensuring that supplies and support were directed to the appropriate company.

Fig. 19–1. The battalion meeting. This team-building session is a forum for brainstorming and problem-solving. (Courtesy Al Pratts)

Before closing the meeting, Battalion Chief Powers made certain that all officers understood their roles and responsibilities. Prior to the date of inspection, company commanders were to make available all inventory reports and maintenance requests. Battalion Chief Powers informed the officers that they would accompany her while the inspection was conducted. She also informed the officers that they would be required to accompany her during any future visits to the firehouses (fig. 19–2).

Battalion Chief Powers then answered any questions and made clear that any issues or concerns were to be brought to her attention. She recognized the work accomplished by her officers and thanked them for their time and participation.

After closing the meeting, Battalion Chief Powers called Division Chief Strong to inform him on the progress made toward completing the order she had been given. The deputy chief thanked her and asked that she supply him with the inspection notice and schedule.

Fig. 19–2. Company officer involvement in the battalion inspection process is essential. (Courtesy Al Pratts)

Battalion Chief Powers's Skills List

Once her officers left, Battalion Chief Powers reviewed what had taken place during the meeting, which she felt had been very productive and, overall, a success. Battalion Chief Powers then wrote a list of the leadership skills employed during the meeting.

The following leadership and management skills and tools were evident in the preinspection meeting:

- Autocratic, participative, and democratic leadership styles
- The management cycle: planning, organizing, implementing, evaluating, remediating, and monitoring
- Problem-solving, direction, communication, and specificity
- Chain of command, unity of command, and span of control
- Delegation, division of labor, and specialization

- Communication: written orders, verbal orders, transactional analysis, and team building
- Maslow's hierarchy of needs: self-esteem, social acceptance, and self-realization
- Evaluation: feedback and self-assessment

After-Action Analysis

- The chain of command and unity of command were adhered to from the top down and the bottom up.
- Both written communication and verbal communication were used to facilitate proper completion of assignments.
- All orders were issued in a clear and concise manner. Feedback was requested, and all orders were acknowledged as understood.
- The dynamics of the formal and informal groups were directed toward achievement of the goals of the assignment.
- The span of control was manageable.
- Planning, scheduling, organizing, monitoring, and remediation were functions evident throughout the assignment.
- Delegation of duties, division of labor, and specialization were used to spread the workload in the fairest and most effective manner.

Final Word

The division chief's inspection assignment was designed to allow Battalion Chief Powers to build confidence and demonstrate command presence. A formal structured meeting created a framework for Powers to exercise control in a safe and supported context emphasizing her command presence (fig. 19–3). Including the company officers in the inspection planning served as a team-building exercise and created a vital group image within the ranks of the officers.

Fig. 19–3. Command presence is needed to set up shop at this job. (Courtesy Ron Jeffers)

chapter 20

CASE STUDY: THE POWERS PREINSPECTION—PART III

Inspection of Ladder 20

Ladder 20 was under command of Captain Pike. At this company's inspection, Battalion Chief Powers discovered that the K-12 saw blade had been put on backward and ordered the condition corrected immediately. She then documented this condition on her checklist. Once the blade was properly replaced, Ladder 20 passed the inspection and was found to be in service and ready.

Inspection of Engine 16/Ladder 22

The next firehouse inspected was a double-company station. Ladder 22 and Engine 16 were located at the southern end of the district. Captain Rung was the ladder officer, and Captain Stretch was in charge of the engine company.

In the course of inspecting Ladder 22's equipment, Battalion Chief Powers asked Rookie Firefighter Tom Jones to start the K-12 saw. Firefighter Jones had successfully completed engine qualification training and was on his sixth week with Ladder 22, working toward ladder qualification. For the first three weeks, Jones had cross-trained but did not ride with the ladder; since then, he had ridden as a ladder company member.

Firefighter Jones appeared nervous and tentative to the battalion chief's eye. Jones had trouble starting the K-12 saw—so much so that

Captain Rung stepped in and started the saw himself. Battalion Chief Powers knew immediately that there was a serious problem in this firehouse. As soon as the saw was properly stowed, Battalion Chief Powers asked to use the company commander's office and ordered for Captain Rung and Captain Stretch to follow her into the office.

Battalion Chief Powers got right to the point. Firefighter Jones's inability to start the K-12 saw was unsafe and therefore represented an unacceptable condition. The first order she gave was to move Firefighter Jones from Ladder 22 to Engine 16; Jones would be replaced by a ladder-qualified firefighter off Engine 16. Rung and Stretch were ordered to go downstairs make the change of assignment, and when the switch was complete, Captains Stretch and Rung reported back to Chief Powers. The company commanders reported that the move was accomplished and that all riding assignments were confirmed. Powers informed the captains that Firefighter Jones would return to cross-training. Powers said she would review all engine and ladder company training records to determine how the issue was to be addressed. Powers then returned to the apparatus floor to complete her inspection.

Engine 16/Ladder 22 subsequently passed inspection and were found to be in service and ready. Before Battalion Chief Powers left the building, she met with Firefighter Jones in the presence of Captain Rung and Captain Stretch. Powers informed Jones that the move to Engine 16 was not a punishment and explained that all companies in her command were to be in service and ready to operate at all times. Powers counseled that it was unfair as well as unsafe for Jones to be riding as a member of a ladder company before he was fully qualified. She also informed Jones that his ladder training would continue and he would return to his ladder assignment as soon as he was fully trained and capable.

After-Action Analysis

- Battalion Chief Powers encountered K-12 saw issues at two separate ladder companies.
- Battalion Chief Powers's primary responsibility was to ensure that all companies were properly staffed, in service, and ready.

Once that was accomplished, she was free to take other appropriate actions.

- The issue with the K-12 saw blade on Ladder 20 had been remedied, and Ladder 20 was in service and ready.
- Firefighter Jones's inability to start the K-12 saw at Engine 16/Ladder 22 created an unsafe condition, which was therefore an unacceptable condition. Jones could not remain on the ladder. Ordering the change of assignments put both Engine 16 and Ladder 22 in service and ready.
- Battalion Chief Powers counseled Firefighter Jones and made it clear to him that the change of assignment was in response to a safety issue and not punitive.
- Problems with K-12 saws in two firehouses required Battalion Chief Powers to find the underlying causes and rectify each situation.

Battalion Chief Powers's Post-Inspection Investigation

Battalion Chief Powers informed all companies to have training reports available for her to pick up during her evening rounds. She then called the division chief to report on the results of her inspection. She reported the actions taken at Ladder 20, Engine 16, and Ladder 22 and briefed Division Chief Strong on the action she would take to remedy the K-12 saw training gap that was identified.

Battalion Chief Powers reviewed all company-level training records. All assigned training was found to have been completed as prescribed and properly documented. Studying the records revealed that the bulk of company-level training depended heavily on demonstrations, and Powers knew from experience that giving a demonstration was not a simple matter.

Battalion Chief Powers reached out to the training officer to determine how much instructor training was given to company officers. The training officer told her that all newly promoted company officers were sent to the academy for a three-week officer training program, part of which covered instructor training. However, no further instructor training was mandated after graduating from the three-week course.

Remedial actions

Battalion Chief Powers determined that the underlying reason for the deficiencies in K-12 saw proficiency had to do with the way that in-service training was being carried out by the company officers. There seemed to be inadequate instructor training for her company officers. She would train her officers how to teach firsthand. Powers would train all of her officers in the proper method for conducting a demonstration (fig. 20–1).

Fig. 20–1. Even informal training sessions need a plan. This is because no fire service training can afford to be informal. Instruction using the demonstration method brings the underlying theory and information found in books right to the students through observation, explanation, demonstration, and practice. (Courtesy Al Pratts)

Instructor training

Battalion Chief Powers conducted instructor training for all company officers. The training focused on the four-step method of instruction and the proper way to conduct a demonstration.

Each company officer was assigned to prepare a demonstration. The battalion chief had the officers choose between SCBA, K-12 saw, or fire extinguisher as the subject of their demonstrations. The company officers were instructed to prepare and present a demonstration to all officers and firefighters in the battalion (fig. 20–2). The officers were ordered to be ready to present instruction beginning on the following tour of duty.

At the conclusion of the training session, Battalion Chief Powers supplied each officer with two handouts. One listed the four steps of the instruction method, and the other was a guide for presenting a demonstration (figs. 20–3 and 20–4).

COMPANY DRILL/DEMONSTRATION
SUBJECT OUTLINE

Platoon: _____ Batt: _____ Company: _____ Date Submitted: _____

Planned Date of Drill: _____ Officer: _____

Drill/Demonstration Subject: _____

Method of Instruction: Classroom   Hands-on   Both

Additional Equipment: Video   Overhead   Slides   Handouts   Test
                          Ladders   Hose   Appliances   Rope

Other
Equipment: _____
_____

Objective: _____
_____

Outline: _____
_____
_____
_____
_____
_____
_____
_____
_____
_____
_____
_____
_____
_____

Subject Approved: ____/____/____   Division Chief: _____

Drill Completed: ____/____/____   Battalion Chief: _____

Fig. 20–2. Company drill demonstration outline. Standardized forms supported by policy allow for consistency across the department.

Handout #1:
The 4-Step Method of Instruction—PPAT
(Prepare, Present, Apply, Test)

Step 1.   Prepare the mind of the learner

_____

_____

Step 2.   Present the information

_____

_____

Step 3.   Have the student apply the information

_____

_____

Step 4.   Test the student's understanding or skill level

_____

_____

Fig. 20–3. Handout on the four-step method of instruction.

Handout #2: Demonstration Checklist

Greet the student(s) _____

Introduce yourself _____

Introduce the lesson _____

Speak about personal experience with the tool or equipment

_____

Explain the benefits of being familiar and proficient with the tool or equipment _____

Present the objectives of the lesson _____

Provide theory, background, and safety information _____

INSTRUCTOR PERFORMS THE DEMONSTRATION

Objective: Explain the purpose and use of the tool _____

Objective: Explain the care and maintenance of the tool _____

Objective: Perform the task at normal speed _____

Objective: Perform the task at reduced speed and explain each step as you go _____

Objective: Perform the task and have the student explain each step

_____

STUDENT PERFORMS DEMONSTRATION

Objective: Have the student perform the task while you guide the student through each step _____

Objective: Have the student perform the task and explain what is being done at each step _____

Objective: Have the student perform the task at normal speed

_____

Objective: Review, test, and summarize the task _____

Fig. 20–4. Handout serving as a guide for demonstration of equipment

chapter 20 | Case Study: The Powers Preinspection—Part III

Battalion Chief Powers's Skills List

Battalion Chief Powers empowered her company officers by providing them with the skills necessary to conduct effective company-level training. Powers's actions would have an immediate impact and long-term benefits.

Once again Powers listed the leadership skills integrated into her inspection and problem-solving actions:

- Investigation, problem recognition, and evaluation
- Problem-solving, requiring coordination between line officers and staff
- Planning, organizing, monitoring, and directing
- Adherence to chain of command and unity of command
- Remediation using positive discipline
- Officer development and empowerment
- Communication: written orders and verbal orders
- Delegation, division of labor, and specialization
- Span of control
- Orders and transactional analysis
- Team building

chapter 21

CASE STUDY: FIRST-DAY EXPERIENCE

You Know It When You See It

The Jane Asher and Johnny Newton case studies presented in this chapter provide the reader with two scenarios describing two very different first-day experiences. Probationary Firefighter Jane Asher has been assigned to Engine 22, and Probationary Firefighter Johnny Newton has been temporarily assigned to Squad 7.

Part I: Probationary Firefighter Jane Asher's Orientation

Scenario

Probationary Firefighter Jane Asher has been assigned to Engine 22. Jane graduated top of her recruit class at the academy. Today is her first tour of duty.

Jane arrived at the fire station 20 minutes early. She is in proper uniform and has her turnout gear, including the face mask she was issued at the academy.

Probationary Firefighter Asher was met at the front door by Captain Rite (fig. 21–1), the company commander of Engine 22. Captain Rite greeted Jane in a friendly manner with a firm handshake: "Welcome to

Engine 22. I know I speak for the entire company when I say that we are all glad to have you join the team."

Captain Rite got right down to business: "You and I are going to kick-start your orientation training right now. We are going to go over the duties you will be responsible for first thing, every time you report for duty. Before we get started, do you have any questions or concerns?" Jane assured the captain that she had no questions and was ready to start her orientation.

Captain Rite referred to a form on his clipboard and began, "Your priority responsibilities must always be attended to as soon as you report for duty. Priority responsibilities are confirming your riding and tool assignments, properly setting up your PPE, and checking your SCBA and PASS device. When all of those items are secured, you will conduct a 360-degree inspection of the apparatus and all compartments. Do you have any questions?" Jane did not.

Fig. 21–1. Primacy means to get it right the first time. The importance and influence of a probationary firefighter's first officer cannot be overstated. Any company officer given the responsibility of mentorship must understand that it is a sacred trust and not to be casually approached. The degree of planning, coaching, and caring that an officer puts into mentoring will stay with a firefighter for their entire career. (Courtesy Brett Dzadik)

chapter 21 | Case Study: First-Day Experience

Captain Rite went on, "I'll show you where and how to place your PPE and give you your riding and tool assignment. I'm going to buddy you up with a veteran firefighter. You will be teamed with Firefighter Earl Grey. When an alarm comes in, follow his lead. Upon arrival at any incident, you are to stay with Firefighter Grey. He is the most experienced man in the company. Do not leave his side and follow his direction."

Captain Rite introduced Jane to her new mentor and added, "I have discussed all of these matters with Firefighter Grey, and he has been fully briefed on his responsibilities and my expectations. Is all of this information clear to you? If you have any questions regarding what we've discussed, please ask." Jane confirmed that she understood.

The veteran firefighter, Earl Grey, extended his hand for her to shake: "Good to meet you, Jane. Welcome to Engine 22."

Captain Rite continued with his instructions: "When your PPE is properly situated, Firefighter Grey will have you perform your SCBA and PASS device checkout. You will also don your SCBA and perform a mask-fit test."

Captain Rite checked off items on his clipboard as he went through the orientation. "If everything checks out, Earl will take you through an initial apparatus familiarization. You will see which compartments hold what equipment. You are not expected to remember everything from one inspection, but it is a key part of your first official training lesson. When all of your priority duties are completed, Firefighter Grey will introduce you around. You'll have some breakfast and meet the rest of the firefighters you will be working with.

"Housework begins at 0900 hours. You will be excused from housework today. You are to report to my office at 0900 hours. I've set aside that time to meet with you. I will discuss your roles and responsibilities as the junior member of Engine 22. I will lay out my expectations. You'll be given an idea of how your first day will unfold. I will also answer any questions you may have. We will meet a second time at 1630 hours to review the day and address any administrative or company-related issues that crop up. I will provide you with a training schedule that covers your next eight tours of duty. Do you have any questions about what we have discussed?"

Again, Jane said that she had no questions and understood what had been explained to her.

Before ending the orientation, Captain Rite referred a final time to the checklist on his clipboard. "Once again, I want to welcome you to Engine 22. This is a very good engine company. The firefighters in this company are some of the best I've ever worked with. You are a welcome addition to this company, and I am confident that you will fit right in. I will see you in my office at 0900 hours" (fig. 21–2).

Fig. 21–2. The most important resource in the fire service: the men and women who serve (Courtesy Brittany Hoffman)

After-action analysis

Captain Rite created a safe, structured, and welcoming environment. Captain Rite tailored his presentation to the level of his audience. This was Jane's first day. The right tone was conveyed; Captain Rite's command presence was powerful but not intimidating. In addition, Captain Rite demonstrated a high level of competency. Competency and command presence are essential to creating a comfortable and trusting bond between supervisor and subordinate.

A checklist and script were prepared ahead of time. These were used to structure and guide the orientation, ensuring that all pertinent information was presented.

The tone used by Captain Rite was firm, fair, and friendly. Information presented was focused, clear, concise, and digestible. Opportunities for Jane to ask questions were built into the presentation.

The primary concern was to ensure Jane's safety as a probationary firefighter. Issuing clear, concise, and direct orders ensured that Firefighter Asher would be in service and ready. Moreover, by making her ready to respond and assigning her to buddy with a veteran firefighter, Captain Rite maintained the in-service and ready status of the entire company. Accommodations for alarm response were addressed. Thus, Firefighter Asher was smoothly and safely integrated into the company.

Overview

The reader is challenged to identify where and how the following leadership skills and tools came into play during Jane's orientation: Planning, scheduling, delegation, command presence, division of labor, specialization, hierarchy of needs, primacy, group dynamics, transactional analysis, "Each one teach one," team building, effective communication, chain of command, and unity of command. All of these were seamlessly integrated into Captain Rite's initial meeting with Probationary Firefighter Jane Asher.

Captain Rite's checklist

Captain Rite recognized that Probationary Firefighter Jane Asher's orientation was a formal training session. His meeting with Firefighter

Asher needed to be planned in advance and delivered in a clear, concise, and comprehensive manner. To achieve these goals, Captain Rite drew from his leadership toolbox and created a lesson plan. The format Captain Rite decided on was a checklist combined with a simple fill-in-the-blank section (fig. 21–3). With this simple preplan, Captain Rite was prepared and able to deliver a comprehensive and successful orientation. The lesson-plan format doubled as a record of the training.

Orientation Checklist

Date: _____ Time: _____

Activity: _____

Participants: _____ _____ _____

Greeting & Introductions: _____

Riding & Tool Assignment: _____

PPE: _____

SCBA: _____ PASS: _____ Mask-Fit Test: _____

Assign Asher to Grey: _____ Outline Alarm Responsibilities: _____

Apparatus Familiarization: _____

Introduction to Firefighters: _____

Schedule 0900 & 1630 Hours Meetings: _____

Allow Time for Questions: _____

Fig. 21–3. Sample fill-in-blank checklist for probationary firefighter orientation

Part II: Probationary Firefighter Johnny Newton's Orientation

Scenario

Probationary Firefighter Johnny Newton has been temporarily reassigned from Engine 15 to Squad 7. Johnny Newton has been on the "JOB" for six months. Today is his first day of specialized training at Squad 7. This case study consists of a brief scenario followed by a short commentary section.

Although Firefighter Johnny Newton is regularly assigned to Engine 15, he has been detailed to Squad 7 for four weeks of specialized decontamination training. Firefighter Newton arrived for duty at Squad 7 half an hour before the start of shift. No one met him at the door, and there was no one on the apparatus floor. Newton laid his turnout gear on the floor and went in search of someone to get some direction from. First, he made his way to the kitchen, where on-duty firefighters and firefighters going home were having coffee, joking around, and swapping war stories. The firefighters warmly welcomed Newton as they introduced themselves. A call was made to Captain Slapdash's office. Firefighter Newton was directed to go up to the captain's office at the top of the stairs.

The office door was open, and Captain Slapdash was speaking on the phone. Firefighter Newton stood in the doorway waiting politely. Only when the captain finished his conversation did he turn his attention to Firefighter Newton. Without rising from his chair, the captain said, "Welcome to Squad Seven. Seven is the best damn company in this whole damn department!"

Captain Slapdash finished up their meeting in short order: "I'll get with you sometime later this morning, okay? In the meantime, you can go back downstairs. Ask the guys where to put your gear and get yourself squared away. We never sweat the small stuff around here, so don't worry about anything. I'm sure you'll pick it up as you go along. Good luck!"

As directed, Newton went back down the stairs to the kitchen.

After-action analysis

There's not much to say. After all, as we've been saying, you know it when you see it!

Johnny Newton's experience reporting for duty was quite different than Jane Asher's. Which of the two experiences would you prefer to have if you were in the boots of the firefighters in the preceding scenarios? Which of the company officers would you model your own actions after? How would you have conducted Johnny Newton's orientation? (See fig. 21–4.)

PROBATIONARY FIREFIGHTER JOURNAL

Probationary Firefighter Name:

Date:	Company:	Captain:

I received training today on:

I read this:

I responded to these alarms and performed the following actions:

I went to these fires and performed the following actions:

I learned the following lessons:

Officer Signature:

Fig. 21–4. NHRFR Probationary Firefighter Journal. NHRFR provides for and requires probationary firefighters to maintain a running diary of probationary training and experiences. Reviewed with the company officer at the end of each tour of duty, the rationale is that awareness requires self-awareness.

chapter 22
CASE STUDY: LADDER 13/32

You Know It When You See It

Firefighters should—and generally do—feel confident that their colleagues will have their back in the hard environment of fires, rescues, and emergencies. The same should hold true in the softer environments. Much too often, casual attitudes, absentee leadership, failure of imagination, and failure to respect the power of the exponential factor combine forces to create unsafe and even life-threatening conditions. The spectrum of negative consequences ranges from minor inconveniences to tragic or catastrophic events.

Apparatus Changeover: Ladder 13 to Ladder 32

Battalion 1's roster was as follows:

On duty

Battalion 1 commander: Battalion Chief Rob Focht

Ladder 13 officer: Captain O. T. Smith

Relieved by

Battalion 1 commander: Battalion Chief Grace Powers

Ladder 13 officer: Captain George Brown

Ladder 13 was backing into quarters at 0130 hours after returning from a false alarm. The headlights and warning lights on Ladder 13 started to blink on and off.

Captain Smith, in command of Ladder 13 at the time, reported the condition to the battalion chief, Rob Focht. Focht ordered Smith to inform dispatch that Ladder 13 would be out of service until it changed over to a reserve ladder.

Battalion Chief Focht next ordered fire dispatch to have another ladder cover Ladder 13's district until the company was placed back in service. The battalion chief contacted the maintenance shops and arranged for reserve Ladder 32 to be delivered and swapped for Ladder 13, and Ladder 32 was delivered to replace Ladder 13 at 0233 hours.

Department policy regarding placing a reserve apparatus into service is comprehensive and specific. Prior to placing any reserve unit into service, a thorough and detailed inspection and inventory were to be conducted. The status of all equipment must be in service and ready.

Captain Smith was tired and anxious to get back his office and his rack. Smith took a quick walk-around Ladder 32. By now it was 0300 hours. After this cursory inspection, Captain Smith called Battalion Chief Focht to report that Ladder 13/32 was in service and ready to roll.

Shift change

At shift change, Battalion Chief Focht was relieved by Battalion Chief Grace Powers. Powers was briefed by Focht regarding all change-of-shift information. Focht informed Powers that Ladder 13 was placed out of service because of an electrical problem. Reserve Ladder 32 was placed in service. Battalion Chief Focht reported that Captain Smith had called at 0300 hours certifying that Ladder 13/32 was in service and ready.

At Ladder 13's quarters, Captain Smith was transferring command of Ladder 13/32 to his relief, Captain Brown. Smith informed Captain Brown about the lights on Ladder 13 and that a reserve apparatus was placed in service in the middle of the night.

Before Captain Smith could complete the transfer of command, an alarm came in, cutting Captain Brown's briefing short. Ladder 13/32 was first to arrive on the scene. Small flames could be seen coming from a litter basket.

Captain Brown radioed in the situation and reported that Ladder 13/32 could handle it with an extinguisher. Dispatch returned all other responding units. Battalion Chief Powers shut off her siren and warning lights but decided to ride by to check things out. Upon arrival, she saw three firefighters in full PPE with SCBA standing by watching a trash-can fire burn brightly. On the ground, next to the firefighters were three extinguishers. An elderly man was coming out of the corner store with a bucket of water, which he used to extinguish the fire.

The firefighters stirred the contents to ensure that the fire was out. The Department of Public Works then arrived, and Ladder 13/32 firefighters grabbed the extinguishers and went back to their truck.

Captain Brown walked over to the battalion chief's car and began to explain. He told her all of the extinguishers were empty. Powers nodded and told Brown to place his company out of service. Battalion Chief Powers informed dispatch to have another company cover Ladder 13/32's district. Captain Brown did not need to be told that the battalion chief would be waiting for him back at quarters.

Remedial actions

Priority one is always to fix the most critical condition first. Battalion Chief Powers witnessed the entire trash-can fire operation. Captain Brown came over to her car. Before he could get a word out, Powers told him to place his company out of service. Brown was ordered to return to quarters. He was instructed to make his company in service and ready to respond. An inspection and inventory were to be conducted posthaste. Powers ordered fire dispatch to cover Ladder 13/32's district with other units until it was back in service and ready.

The battalion chief recognized that Ladder 13/32 was not 100% in service and ready. If the extinguishers were not serviceable, then what else might not be operational on the reserve ladder? This was an unsafe condition—and therefore unacceptable.

Battalion Chief Powers would not allow herself to fall victim to wishful thinking. Hoping that the dead extinguishers were the only problems and hoping that the reserve apparatus was otherwise service ready was not an acceptable option. Battalion Chief Powers's responsibility was to ensure the safety of personnel and provide fire, rescue, and emergency response to the community. Thus, she would have to ensure and know that all apparatus and equipment in her command were able to meet those responsibilities.

The battalion chief is responsible for having all companies under her command certified as in service and ready. Whenever there is any question regarding life safety or protection of property (or both), the mandate is to err on the side of safety.

Once the life safety and operational readiness of the reserve ladder were certified as in service and ready to respond, all other issues could be addressed. Battalion Chief Powers met with Captain Brown as soon as he reported that the reserve ladder was back in service and ready.

Captain Brown explained that he had been told that the status of Ladder 13/32 was in service and ready. Brown told Battalion Chief Powers that the alarm came midway through the transfer of command. In response, Powers told Brown to submit a written report detailing the situation from shift change through to certifying the reserve apparatus as in service and ready. Finally, before Captain Brown was dismissed, Powers asked him, "Which piece of equipment on your ladder is the most important piece of equipment?"

Brown thought for a moment, then replied, "The SCBA."

Powers shook her head. "No, Cap, it's not the SCBA."

Brown looked flustered and puzzled. "Well, Chief, what is the most important piece of equipment on the ladder?"

Powers replied, "The most important piece of equipment on any engine or ladder is the piece of equipment that is needed at that moment" (fig. 22–1).

Department policies and procedures worked as intended. A maintenance issue was discovered, the condition was reported up the chain of command to the battalion commander, and remedial action was initiated. The battalion chief coordinated between line and staff personnel to have a spare unit delivered to replace the out of service rig.

Fig. 22–1. This is the last place you want to find out that in-service and ready status of the apparatus is compromised. (Courtesy Ron Jeffers)

Overview

The formal organization had in place an SOP for proper changeover of apparatus. The prime focus of the changeover procedure is to ensure the life safety of operating members and the in-service and ready status of apparatus and equipment.

The integrity of the life safety and the in-service and ready status both were compromised by Captain Smith's failure to conduct a thorough inventory and inspection. Smith's negligence was compounded when he misrepresented the serviceability of the replacement ladder truck. Calling the battalion chief and falsely reporting on the status of the reserve ladder truck was a gross violation of trust in blatant disregard for the safety and well-being of all other firefighters. Ladder 13/32 caught an alarm during shift change. Captain Brown was responding on a spare ladder that was improperly cleared as in service and ready. At the time when Captain Brown radioed that Ladder 13/32 could handle the situation, that assessment was based on information that was incomplete and erroneous.

Per procedure, firefighters brought two water extinguishers to extinguish the small trash fire. Neither of these two water extinguishers was functional, though. Captain Brown next called for the dry-chemical extinguisher, but the dry-chemical extinguisher was not charged and thus was of no use. At about this time, a civilian walked up with a bucket of water and dumped it onto the burning contents of the trash can.

The failure to follow an SOP resulted in Ladder 13/32's responding with an incomplete inventory of service-ready equipment. Even though, on arrival, Captain Brown had performed a proper size-up, his knowledge of the equipment was inaccurate, and his decisions and directions reflected that. The equipment necessary to safely and effectively mitigate the situation was not functioning. The ladder company was left to stand and watch in embarrassment as the fire was extinguished by a civilian instead.

Individually, any of the errors described in this scenario might be considered no big deal. However, there are very few truly minor moments in the fire business. Something that may seem to be "no big deal" at one moment is a potential point of failure at another (fig. 22–2). The reader is challenged to come up with a way to deal with Captain Smith's false reporting.

Fig. 22–2. The most important piece of equipment is what is needed right now. From the looks of this situation, that ground ladder looks like it is pretty high on the needs list. (Courtesy Ron Jeffers)

Final Word

The impact and consequences of this incident extended upward and outward. Battalion Chief Powers would submit a report to her division chief. Then, the division chief would be required to forward any reports of unsafe conditions to the chief of department.

Unintended Consequences

A wrinkle morphs into an unintended publicity consequence

A video recording of the trash-can fire, including the civilian involvement, was posted to YouTube and circulated through social media. The video was entitled "Candy store owner rescues three firefighters from a trash fire."

There is no need to stretch this analysis any further. By now, the reader has witnessed a seemingly insignificant failure to comply result in an enormous expenditure of valuable time, energy, and person-hours.

Postscript

While browsing a sidewalk craft fair I happened upon a small wooden sign. The sign read, "The greatest shortcoming of the human race is our inability to understand the exponential function." At the time, I did not have a good idea of what was meant by this quotation, of the physicist Albert Allen Bartlett. However, I do now. The actions and consequences described in the preceding case study provide a small lesson regarding the awesome power of the exponential factor.

chapter 23

CASE STUDY: THE MYTH OF THE GREAT FIREFIGHTER

You Know It When You See It

"Yeah, but he's a great firefighter." This statement seems, on its surface, to be a high compliment, declaring that a particular firefighter works well—or better than well—on the fire, rescue, or emergency ground. When the term "great firefighter" is preceded by "but," it is often a line in story about a firefighter who was involved in some sort of unacceptable behavior.

In my experience, "Yeah, but . . ." is usually the prologue to excuse making, plea bargaining, blame shifting, cover-up, or outright mendacity. "Yeah, but . . ." is also code for "Come on, Chief. Give the guy a break." This mentality gives license to behavior that is easily recognized as inappropriate.

In my 30-year career, I have had the privilege to work with truly great firefighters. These were great firefighters with no if, ands, or buts about them. From these colleagues, I learned that a great firefighter is a full-time position. A great firefighter is being right there and doing the right thing in both the hard and soft environments. A great firefighter is a 360-degree fire, rescue, and emergency operator—and a 360-degree firefighter is a positive force and example in and out of the firehouse. A great firefighter is a lifelong student and teacher of the art, science, and traditions of the fire service. Those great firefighters also taught me that a good firefighter does not hide behind "Yeah, but . . ." and that a good boss does not make excuses—a good boss makes things right.

These same firefighters told me that I would not always get it right. In my case, the general consensus was that I would more often get it wrong than I would ever get it right. But that didn't matter! My job was to aspire to be and work at becoming the best firefighter I could possibly be. Great firefighters are not always great, but they always aspire to greatness.

A great firefighter has your back and, at least as important, has the courage of conviction to tell you when to back off and when you are dead wrong. Men and women focused on doing the very best they can under the most extreme circumstances is an apt definition for the fire service.

Something Real That Might Have Happened

Hazing. *The practice of rituals and other activities involving harassment, abuse, or humiliation particularly when used to initiate a person into a group.*

Bullying. *The use of force, threat, or coercion to abuse, intimidate, or aggressively dominate others. This often repeated and habitual, and imbalance of social or physical power is invariably behind this behavior.*

Scenario

Engine 40, Rescue 1, and Battalion 5 all ride out of a single station, which is located dead center of the 5th Battalion. This the busiest battalion in the 3rd Division.

Engine 40 and Rescue 1 each ride and respond with an officer and three firefighters. The battalion chief drives separately.

The firefighters and officers, though young, are seasoned and proud. They think of themselves as worthy fire operators. As most firefighters do, they put great cachet in how a firefighter performs on the fireground.

Three weeks earlier, Probationary Firefighter Alex Rivera graduated from the academy. Firefighter Rivera, who is a recently discharged and decorated veteran of the war in Afghanistan, has been assigned to Engine 40 for company-level training. No one is sure if the probationary firefighter would be permanently assigned to Engine 40, although this was assumed to be likely.

From the first day when Firefighter Rivera arrived, Firefighter Tom Wood got on his case in a blatantly inappropriate manner. Firefighter Wood was not at all circumspect about his behavior, and every firefighter and officer was witness to Tom's "antics."

Everyone in the station was used to Tom's abrasive personality. In the past, whenever Tom had tried his bluster out on any of the other firefighters, he was quickly directed to take his attitude and appropriately situate it somewhere very hot and dark, and that would be the end of it. Thus, at first, Wood's behavior toward Rivera was shrugged off as hazing of the probie. However, Tom's behavior toward Rivera was so mean-spirited that it made everyone in the firehouse uncomfortable. The other firefighters felt bad for Alex, but for a long time, nobody—officer or firefighter—said a thing. What had seemed to start as typical hazing had become obvious for what it was—outright bullying. This behavior unfolded in full view of two captains, a battalion chief, and all the firefighters in the house.

Commentary

What was happening at Engine 40's quarters was a clear case of bullying. Bullying is rarely just a top-down exercise—in an organization, it may be side to side or even from the bottom up. Left unchecked, the consequences of bullying radiate in every direction, and this behavior will have a negative impact on every member of the firehouse.

The only reason Tom Wood got away with bullying was that the battalion chief and the two captains were cowed into absentee leadership roles. Confronting a bully is uncomfortable. Not one of the officers stepped up and out of their comfort zone to do that part of their job. In this case, their job was to ensure and maintain a safe and stable work environment. If all or any of the officers had done their job, then the probationary firefighter would have been protected from being abused. Further, the officers could have used this as an opportunity to realign Wood's thinking, attitudes, and behavior with the requirements of the "JOB."

No Leader, No Follower

Scenario (continued)

Firefighters will take their cue from their leaders. In this scenario, because there was no leader intervention, there was no follower intervention either.

My philosophy has always been that the probationary firefighter is the most important member of a crew. This is because the probationary firefighters are the future of the profession.

In this case study, an informal leader not only bullied Rivera but also intimidated the battalion chief, both company officers, and the other firefighters in the station. Fire officers and firefighters had allowed the most junior man to be harassed and abused. An informal leader was allowed to hijack control of the firehouse.

No one should ever try to soften the view of such behavior by assuming it is "just hazing." To do so is to condone bullying, which is never acceptable.

This situation at Engine 40 would eventually escalate and could only end badly. It would have been clear to anyone who was paying attention that Firefighter Wood's behavior was bullying. His actions were harmful and violated the code of conduct outlined in the department's rules and regulations. Even though Tom Wood's bullying was just plain wrong, no officer stepped in to stop him; this is a classic case of absentee leadership.

At breakfast one morning, Firefighter Wood told Probationary Firefighter Rivera that he was sitting in the seat that Wood considered to be "his" place at the table. Wood further explained that if Rivera ever sat there again there would be consequences. Wood then told Rivera to make the coffee, mop the kitchen floor, and then clean the bathroom.

All of this transpired while the battalion chief and the company officers were sitting at the kitchen table. Nonetheless, their conversations went on without a hiccup.

chapter 23 | Case Study: The Myth of the Great Firefighter

At 1000 hours, the rescue and engine left quarters to report for a scheduled training exercise. Firefighter Rivera never got around to cleaning the bathroom. When the companies returned to quarters, Firefighter Wood saw that the bathroom mop and bucket were unused and took it as a personal affront. He decided to teach the probie another lesson. Tom Wood took the wet mop and laid it on Rivera's bed.

Commentary

Tom's thinking relied on twisted logic and was clearly absurd. However, failure to supervise gave license to Tom's distorted worldview. By not intervening, the captains and the battalion chief tacitly endorsed Wood's behavior.

Discovering the mop on his bed, Rivera could no longer suppress the built-up resentment that had been brewing and exploded: "Don't you know this is the kind of stuff that gets people shot?! I'm going to kill you!"

One of the firefighters was able to calm Rivera down before the battalion chief ordered one of the captains to go see what the ruckus was all about, and the captain reported that the situation was taken care of. Thus, no incident report regarding this situation was filed.

This should be viewed as a cautionary tale.

> *Not even a bullet can catch the spoken word*
> *in a fire department.*
>
> —Flood

Now Deputy Chief Grace Powers showed up at Engine 40/Rescue 1 while making her daily rounds. When she arrived at the quarters of Engine 40/Rescue 1, Deputy Chief Powers informed the battalion chief that she had heard about a disturbing situation at Engine 40 while at another fire station.

Deputy Chief Powers was told that Probationary Firefighter Rivera had threatened the life of another member. The battalion chief explained the situation to Deputy Chief Powers and tried to play it off as rookie hazing that had gone a little too far. The deputy chief continued to question the battalion chief regarding the buildup to the outburst.

Once the deputy chief had a full picture of the goings-on, she ordered a report from the battalion chief and both captains detailing the incident, the history of the bullying, and their failure to properly supervise. Division Chief Powers told the battalion chief that he should know that any member threatening another member with grievous and bodily harm was considered to be in gross violation of the rules and regulations of the department. Rivera's threat was unacceptable and a chargeable offense.

Deputy Chief Powers ordered that Firefighter Rivera report to her in the battalion chief's office. Powers wanted to interview Probationary Firefighter Rivera to get his side of the story and ascertain his state of mind and his ability to perform his duties. Powers was satisfied that Rivera was composed and capable to perform his assigned duties.

Next she interviewed Firefighter Wood. Both Wood and Rivera were required to submit reports explaining their involvement in the incident.

Finally, Division Chief Powers called both captains and the battalion chief into the office for a meeting. The division chief asked why Wood's behavior had been tolerated for so long. A number of weak and lame responses were offered. The captain of Engine 40 ended the meeting by saying, "Yeah, but Wood is an great firefighter." The division chief just shook her head in dismay.

Commentary

Where to begin commentary of this total abdication of command may be the most difficult decision. The amount of supervision, management, and leadership exhibited by the officers in the scenario could fit into a thimble and still leave room for a 35-foot ladder!

Two investigations were required for the deputy chief to unravel all that had transpired. The first investigation started at the end and worked back to the beginning, and the second investigation started from the beginning and moved through in sequence to the end.

In situations like the one described, it is easy—and, therefore, a tendency—to focus on the participants, namely Wood and Rivera in this scenario. It is even easier to focus only on Rivera, and that is exactly how this scenario was resolved in real life. Rivera was brought up on charges and lost two vacation days.

chapter 23 | Case Study: The Myth of the Great Firefighter

The situation had escalated exponentially and could no longer be ignored. At this point, the formal organization was forced to exact its pound of flesh. Thus, a penalty was imposed on Probationary Firefighter Rivera. The formal organization lifted the rug, and the officers involved hoped to sweep the incident right under it. However, hoping problems will just go away or fix themselves is wishful thinking.

The representatives of the formal organization were absentee leaders. The officers created a vacuum that was predictably filled by the informal group.

Hazing and bullying are unacceptable and should be stopped immediately, in all cases. By failing to intervene and stop Wood's behavior, all three officers were sending a message that his behavior was in fact acceptable. When unacceptable behavior is allowed to go unchecked, the absentee leader will end up with a fragmented fire company.

What happens in the firehouse finds its way to the fireground. If a formal leader cannot maintain safety and ensure company integrity in the firehouse kitchen, then the ability to operate safely as a cohesive unit during fire or emergency operations is questionable.

Firefighter Wood did not start running the show in Engine 40's firehouse on the day the probationary firefighter arrived. Informal leaders harvest permission, often tacitly, from formal leaders. This unauthorized influence is accumulated over time and ultimately erodes command confidence and dilutes command presence in every environment, soft or hard. An informal leader "gone rogue" pushes against the envelope of formal authority at every opportunity. What is permitted is promoted.

There comes a tipping point where an officer's ability to confront this behavior dissolves into a state of paralysis. It is easier to stop undesirable behavior quickly. Stopping unacceptable activities immediately not only is easier but also is a formal leader's mandated responsibility. Absentee leaders are regularly playing catch-up—and catch-up is a place at the back of the line, while leadership is the front of the line.

In cases like the scenario described in this chapter, a catastrophic incident (here, the threat of bodily harm) is usually required to initiate the realignment of power in the organization. This will unnecessarily cat up time, energy, and usually money to redress the situation created by the absentee leader. In this case, the absentee leader, having no influence of his own and being powerless as a babe in the woods, had to pass the problem up the chain of command before it could be resolved.

After-Action Analysis

The chief of department required that the division chief provide complete reports from all officers involved. Division Chief Powers was also required to submit a report. All reports were expected to be completed and forwarded to the chief of department by the end of the day.

The division chief was expected to investigate and detail every point in the chain of command that had failed. The division chief was also to recommend appropriate discipline options for all officers involved. The division chief submitted all the required reports along with investigation findings and recommendations for discipline.

The two captains, the battalion chief, and the division chief were called to the office of the chief of department. The chief of department told the officers that he had reviewed all the reports submitted and then told the captains and the battalion chief that each would be suspended without pay for two days. They were to return to his office with their respective union representatives. The chief of department would have the charges for failure to supervise when they returned.

Postscript

The chief of department was told a number of times during the investigation, "But Chief, Wood is an excellent firefighter. We figured it was just squabbling between firefighters and it would blow over."

Before the officers were dismissed the chief of department offered the following summary:

> *To be a great firefighter is a 360-degree commitment.*
> *On this job, 10% of the work is at emergencies and fires.*
> *The other 90% is nonemergency activity. Greatness is a high bar to reach. Very few do. So, a guy like Wood may be good or above average 10% of the time and a negative force 90% of the time. You gentlemen are dismissed.*

OUR LAST WORD(S): CHIEF FLOOD'S PERSPECTIVE PARABLE

Once upon a time, there sat a firefighter on the bank of a gently flowing river. On the shore across from the firefighter and the gently flowing river stood a lush, verdant forest. The firefighter sat serenely upon the shore taking in the view of the gently flowing river and the lush, verdant forest, and to the firefighter, all the world seemed right.

Behind the carefree firefighter, perched upon a large rock, was a company officer. The company officer could see the carefree firefighter, the gently flowing river and the lush, verdant forest, and he could see shadows moving deep in the forest.

Upon a hill, above the company officer watching over the firefighter, stood the battalion chief. From the vantage of the hill, the battalion chief was able to see the captain on the large rock, the firefighter on the shore, the gently flowing river, the lush, verdant forest, and the shadows moving deep in the forest. Across the river, deep in the forest, beyond the shadows, the battalion chief could also hear a distant, ominous rumble.

On a plateau above the hill, farther back from the large rock, towering over the shoreline, the gently flowing river, the lush, verdant forest, the shadows, and the distant, ominous rumble, stood the deputy chief. The deputy chief could see the battalion chief on the hill, the captain on the rock, the firefighter on the riverbank, the gently flowing river, the lush, verdant forest, and the shadows, and he could hear the distant, ominous rumble. Beyond all that, he could see a dark and foreboding cloud of dust rising above the tree line at the far edge of the forest.

On a mountaintop above the deputy chief on the plateau, far back from the battalion chief on the hill, farther back and above the captain on the rock, and towering over the firefighter on the shore, watching the gently flowing river, enjoying the view of the lush, verdant forest, with its shadows, ominous rumblings, and looming cloud of dust, stood the chief of department. From the mountaintop, the chief could see the deputy chief on a plateau, the battalion chief on a hill, the captain on a rock, and the firefighter sitting on the shore.

From this highest of vantage points, the chief could see a hoard of 7,000 Macedonians dressed in full battle array, armed to the teeth, mounted on powerful warhorses, and raising a massive cloud of dust as they entered the forest intent on crossing the river and wreaking havoc and mayhem in and upon every division, battalion, and company within his department.

See?

INDEX

A

absentee leadership
    bullying condoned by, 247–248
    discipline for, 252
    energy investment drained by, 117–118
    productivity blocked by, 117
    symptom of, 121
accountability, xiii, 2, 155
    discipline regarding, 93, 115
    of fire officer, 122
    weak-kneed leader failure of, 21
achievement, xxxix
action
    authority undermined by, 88
    follow-up, 208
    on grievances, 134
    leader inspiring, xl
    loyalty as, 2
    remedial, 220, 239–241
    reverence for, 102
    situation-specific, 17
administration, 14
adult ego state, 86
Agostini, Bobby, xxxiv
Allen, Dan, 68
Allen, Woody, 131
ambush leadership, 126
    expectations confused by, 125
    productivity absence of, 124
analysis, xviii
apathy, 110, 112
apparatus
    changeover of, 237–243
    demonstrations handout for, 224
    inspection of, 128, 137–138
    remedial actions regarding, 239–241
    shift change regarding, 238–239
approval, personal, 65

Aristotle, 129, 167
The Art of War (Sun Tzu), 20
Ascolese, Vincent, xxxvi–xxxvii
Asher, Jane, 227–232, 234
assignments, work
    disregard for, 127
    evaluation of, 208–209
    for inspection, 206–208, 209
A-Team, 145
attention
    capacity for, 14
    communication requiring, 74
    to detail, 9, 12
    for grievances, 134
    need for, 186
attitude
    casual, 158–159
    Dent problem with, 180–185, 186
    improvement of, 179
authority
    absentee leadership failure of, 117
    actions undermining, 88
    delegation regarding, 149, 150–152
    framework reinforcing, 211
    inappropriate behavior reported to, 96
    problems communicated to, 42
    temperance for, 60
autocratic leadership
    alternative styles integrated into, 27
    as authoritarian leadership, 25
    problems with, 27–28
    scenarios requiring, 26, 67
    standardized operating procedures codifying, 68
Avillo, Anthony, xxxvii, xliv–xlv
Avillo, Anthony D., xxxv
Ayivor, Israelmore, 91

B

barrel, of waste oil, 128–129
Barreres, Rich, xxx–xxxi
Bartlett, Albert Allen, 243
behavior, inappropriate
    acknowledgment of, 195, 197
    bullying as, 246–247
    employee realignment of, 95–96, 179
    excuses about, 245
    Finnegan illuminating, 195–196
    report for, 96
    timing stopping, 251
    toleration of, 248–250
beliefs, about leadership, 15–16
Bell, Lawrence Dale, 1
Belsky, Scott, 12
benchmarks, 152, 153
Bergen County Fire Academy, xxxviii
biological needs, 36
Blanchard, Ken, 37
body language, 86
boundaries, 54, 204
    counseling challenge with, 175
    soft environment blurring, 85
Brannaman, Buck, 91
break time, 136–137
Browne, George, 163
budget, xlviii, 31
Bulfinch, Thomas, 37
bullying, 21
    absentee leadership condoning, 247–248
    discipline regarding, 252
    as inappropriate behavior, 246–247
    Powers intervention regarding, 249–250
    ramifications of, 247, 250–251

C

camaraderie, 45–46
captains
    Barreres as, xxxi
    injury of, 105–106
    resentment between, 127–129
career
    command inconsistencies jeopardizing, 5
    education throughout, 97–98, 169, 206, 245
    leadership study through, xli, 33
casual leadership, 109
    leadership void from, 111
    sensual leadership missing from, 110
    tragedies from, 114
character, 7
chief
    apathy result of, 112
    battalion and deputy, 253
    competency responsibility of, 70, 118
    Constantinople potential for, 132
    interpersonal conflict with, 87–89
    operations, xxv–xxvi
    planning lacking by, 128–129
    regulation enforcement of, 120–121
    reinforcement by, 122
    as role model, 118–119
    as servant-leader, 113
    view of, 254
child ego state, 86
Churchill, Winston, 158
Cicero, Marcus Tullius, 1
coaching, 170, 171
    by Avillo, Anthony D., xxxv
    communication exercise of, 173, 199
    as confrontation, 200
    education on, 184–185
    between Finnegan and Slip, 193–198
    planning required by, 173
    rules for, 176
    for software, 172
coercion, 59
cognition, 102–103
cognitive premeditation, 102–103
collaboration, 8, 28–29
comfort zone, 45
    interpersonal conflict resolution outside of, 47
    leader looking beyond, 51–52
    promotion leaving, 163–164
    responsibility over, 53
    subordinates regarding, 38
command
    abdication of, 250
    balance for, 66
    career impacted by, 5
    chain of, 70, 118, 188, 189
    cognition challenges for, 102
    comfort in, 50

flexibility for, 150, 151
hard environment communication enabling, 80
image, 9
philosophy, 159–160, 164–165, 191–192
presence, 164, 177, 200, 206, 208, 215
responsibility of, 204
situations threatening, 51
technology enhancing, 79
total-immersion leadership for, 67
trust gained from, 164
unification of, 52
vacuum, 21
verbal and nonverbal, 26
of Weehawken Fire Department, 16
commanders
empowerment of, 150
ownership of, 204, 209
prerogative of, 198
responsibility understanding of, 205
communication, xxiii
clarity of, 43
coaching exercise in, 173, 199
comprehension required for, 73, 75
counseling as, 174, 176, 179, 199
for delegation, 152–153
discipline product of, 97–98
within education, 97–98
of expectations, 157–158
failure of, 107
fire service leader scrutiny for, 81–84
in hard environment, 80–83
informal group countered by, 5
by intellectually curious leaders, 13
meetings opening, 162, 206
nonverbal, 77, 177
by Powers, 212, 214, 219
practice for, 82
regarding problem-solving, 42, 84, 88–89
role model of, 88–89
skill portfolio for, 83, 87
in soft environment, 81–83, 84–85
strategy for, 74, 81
top-down and bottom-up, 139
verbal, 76, 77

of vision, 6, 159–160
visual, 78–79
written, 78
community
department investment by, 96
education of, 156
expectations of, 97
company, 222
company commander, 127
competency
chief responsibility for, 70, 118
through command chain, 70
for communication, 74
leadership failure for, 148
obedience measuring, 8
weak-kneed leader compromising, 22
complaints, 40, 134
comprehension, xviii
communication requiring, 73, 75
of emotional interactions, 33
of expectations, 157
of leadership requirements, 178
compromise, 49
conduct, code of, 155
confidence
through discomfort, 50–51
from interpersonal conflict resolution, 49
Powers' challenge with, 203, 209, 211, 214
conflict, interpersonal
confidence handling, 49
in fire companies, 46–47
in interpersonal communication, 76
over kitchen rack, 87–89
leadership states for, 48
timing of, 50, 52
confrontation
coaching and counseling as, 200
honesty requiring, 6–7
transfer avoiding, 181
weak-kneed leader avoidance of, 21–22
connection power, 64
consistency, 5–6
Constantinople, Frank, xxx, 132–133
contingencies, 12
contract
leadership as, 16
oath as, 1, 33
psychological, 138

coordination, xlviii, 12, 31
counseling
    as communication, 174, 176, 179, 199
    components of, 176–178
    conclusion for, 197–198
    as confrontation, 200
    education on, 184–185
    between Finnegan and Slip, 193–198
    guidelines for, 176, 178–179, 193, 201
    hypervigilance for, 175
    skill honing for, 170, 176
    for software, 172
    work environment for, 177
Covey, Stephen R., 73, 120
creativity, 29–30, 141–142
crew comfort, 21–22
culture, 54, 69–70
cursive power, 63
Curtis, Dave, xxxii

D

dabbler, 124
data, 74, 80
deadline, 128–129
decision-making, xxi
    autocratic leadership for, 25–26, 67
    democratic leadership for, 30
    emotional maturity for, 7
    preparation for, 81
    for transfer, 181, 183–184
delegation
    benchmarks for, 152, 153
    command flexibility for, 150, 151
    guidelines for, 152–153
    progression of, 148–149
    willingness for, 147
democratic leadership, 30–31
demonstrations
    apparatus handout for, 224
    company officer training regarding, 219–221
    company outline for, 222
Dent, Stew
    attitude problem with, 180–185, 186
    leadership failure for, 182–183
    maintenance of, 185–186
    transfer of, 181

department, 16, 132. *See also* Weehawken Fire Department
    chief observation of, 120
    community investment in, 96
    company officer bedrock of, 71
    fire company cornerstone of, 45
    formal leader representing, 35, 200
    as formal organization, 3–5
    leader development support from, 109
    leadership asset to, xx
    morale cultivation by, 133
    negative discipline preserving, 95
    ready status expected of, 161
    reputation of, 53, 97, 99, 243
    revitalization of, xxx
    safety responsibility of, 122
    unification of, 156
    weak-kneed leader symptom of, 22
detail, attention to, 9, 12
development, of leaders, xiv, xxxix, 17, 39, 149
    department support of, 109
    discipline education for, 92
    energy investment regarding, xx, 70, 186–187
    for generation shift, xvii, xxi, 28, 98
    nonlinear, xliii
    opportunities for, 205
    program for, 122
    socialized power orientation enhancing, 61
development, subordinate, 18, 39, 61
    delegation for, 147–148
    discipline for, 98
    by leader-coach, 172–173
    leaders responsible for, 148
    personalized motivation preventing, 59
deviation amplification, 104
Dillard, Annie, 12
discipline. *See also* negative discipline
    for absentee leadership, 252
    accountability regarding, 93, 115
    as communication product, 97–98
    as education, 91–92
    within leadership development program, 122
    positive, 93–95, 188
    protocols for, 96–97

Index

of self, 93
as tool, 91–92, 94, 99
tragedies prevented by, 53
transfer compared to, 188
value of, 92
drill session safety checklist, 11
Drucker, Peter F., 37, 63
duty, tour of, 158–159, 169

E

education
  career-long, 97–98, 169, 206, 245
  on coaching and counseling, 184–185
  communication within, 97–98
  of community, 156
  delegation for, 149–150
  discipline as, 91–92, 188
  fire science course for, 132
  instructor training as, 221
  from Malley, xxxvii–xxxviii
  on morale, 133–134
Eisenhower, Dwight David, 57, 131
emergencies
  adaptation for, 169
  communication importance during, 82
  focus for, 124
  information delivery regarding, 80–81
  population service for, xxvii–xxviii
  resource drain regarding, 83
  standardized operating procedures addressing, 72
employees
  expert and referent power influencing, 65
  grievance attention for, 134
  Hawthorne experiments on, 136–140
  improvement of, xxiii
  inappropriate behavior realignment of, 95–96, 179
  motivation study on, 140–143
  recognition influencing, 138, 139
empowerment
  of commanders, 150
  of company officer, 225
  of subordinates, 19, 149–150
Engine 40, 246–249
Engine 16/Ladder 22, 217–218

Engine 22, 217–218, 227–232
Engine 203, xxxviii–xxxix, 191
environment, work. *See also* hard environment; soft environment
  changes in, 136–137, 195
  chief presence in, 119
  control over, 125
  for counseling, 177
  discipline maintaining, 92, 97
  group dynamics influencing, 138
  maintenance of, 47, 49, 172
  morale impacting, 135–136, 140–143
  motivation impacting, 140–143, 153
  for orientation, 231
  power exploitation metastasizing, 57–58
  productivity impacted by, 136–137
  socialized power orientation improving, 61
  synergy in, 167
espirit de corps, 135
Von Essen, Thomas, xiv, xv
Ewing, Russell H., 124
exam, promotional, 68–69, 180–185, 186
excuses, 22, 245
expectations
  ambush leader confusing, 125
  chief setting, 112, 120
  clarity for, 71, 92
  communication of, 157–158
  of community, 97
  comprehension of, 157
  for delegation, 153
  for department, 161
  enforcement of, 156
  for 1st Platoon, 164
  of formal organization, 93, 94–95
  for inspection assignment, 209
  for subordinates, 155
experimentation, 47
expert power, 63–64, 65
experts
  evolution into, 149
  group members as, 29–30
  leaders rather than, 39
exponential factor, 112, 203, 237, 243

F

facilitator, 28, 30
failure, leadership
    of formal organizations, 118
    freelancing as, 114
    without standardized operating procedures, 105
    subordinate incompetency from, 148
    of supervision, 47, 107, 112, 121, 249
    transfer as, 182–183, 189
    victimization from, 106
fairness, 7–8
family, xxxv
    Ascolese analogy for, xxxvi
    fire companies as, 46
    service running in, xxxi
Fighting 1st, xxxvii, 163
Finnegan, Tim, xxxi–xxxii
    counseling session held by, 193–198
    inappropriate behavior illuminated by, 195–196
    meeting held by, 191–192
    as mentor, xxxiii
    recognition by, 193–194, 198
fire company
    changes in, 194
    as department cornerstone, 45
    as family, 46
    interpersonal conflict in, 46–47
Fire Department of New York (FDNY), xiii
fire extinguishers, 239, 242
fire service leader
    communication scrutiny of, 81–84
    generations of, 167, 168
    involvement of, xix
    as linchpin, xx
    parents influence for, xxxv
    presence of, 123
    responsibility of, xliii
    subordinate's perceptions defining, 117
firefighters. *See also* probationary firefighter
    casual attitude of, 158–159
    framework for, 109
    inappropriate behavior excuses of, 245
    inexperienced, 107
    maintenance of, 185–186
    360-degree, 245–246
    veteran buddy system for, 229, 231
    view of, 253
fireground, 84–85
Fireground Strategies (Avillo, Anthony), xxxvii, xliv–xlv
1st Platoon, 163–165
foam extinguishing agents, 150
football, xxxvi, xxxviii
Ford, Henry, xvii
forest, view of, 253–254
formal leader
    contact point with, 40
    as department representative, 35, 200
    informal leader permitted by, 251
    leadership void of, 111, 248
    status quo of, 34
Formal Organization Boulevard, 5
formal power, 62–63
freelancing, 114–115
Friday, Constance Chuks, 91
functional fixity, 104

G

Ganci, Peter, xiv
Gandhi, Mahatma, xvii
generations, of leaders, 140
    development of, xvii, xxi, 28, 98
    in fire service, 167, 168
gloves, 180–181
goals
    Ascolese wisdom on, xxxvi
    creative leadership for, 142
    deviation amplification from, 104
    fire officer and department balance of, 16
    labor division for, 149
    motivation for, 141
    power misuse sabotaging, 57
    of troop return, xlii, 3
good-cop/bad-cop, xxx
Greatest Generation, 140
Greenleaf, Robert K., 20
grievances, 134
Ground Zero, xxxiv
group dynamics, 34, 138, 184, 214
guidance, leadership
    for Powers, 204
    by Strong, 208–209

Index 261

H

hand-signal system, 105
hard environment, xliv
    communication in, 80–83
    discipline during, 115
    interpersonal conflict in, 50
    mind-sets, xlv
    soft environment impacting, 68, 113, 205
    soft-think in, xlviii
    360-degree firefighter in, 245
hard-think
    dark matter compared to, xlv
    for soft environment, xlvii–xlviii
    in training, xlvi
hardware
    importance of, 240, 242
    maintenance of, 170–171, 185
Harry, Deborah, 191
Hawthorne experiments
    motivation factors revealed by, 137–140
    paradigm shift from, 136
hazing, 246–247
Hern, Mike, xxxi, 75
hierarchy, of punishment, 95
high rise operations, 27
Holtz, Lou, 131
honesty, 6–7
Huelbig, Gerald, xxx
human assets. *See* software
human condition, xxi, xxiii, 19
human relations, 6, 170, 200
Humes, James, 85

I

imagination, failure of, 104, 107
imprimatur, xv
improvisation, 17, 81, 85
influence, xl
    direction of, 57
    expansion of, 184–185
    for fire service leader, xxxv
    of informal leader, 35
    on profession, 167
    from rank, xlii
    recognition enhancing, 138, 139
    from referent power, 65
Informal Group Freeway, 5
informal leader, 35
    formal leader permitting, 251
    leadership void filled by, 111, 248
informal power, 63, 65
information
    autocratic leadership based on, 27
    availability of, 5, 241–242
    communication exchanging, 74
    emergency delivery of, 80–81
informational power, 64–65
insecurity, 59
inspection
    of apparatus and station, 128, 137–138
    assignment, 206–208, 209
    checklist for, 207
    compromised, 241
    of Engine 16/Ladder 22, 217–218, 227–232
    investigation after, 219–221
    of Ladder 20, 217
    leadership skills for, 225
    of portable radios, 120, 121
    by Powers, 217–218, 219–221
    preparation before, 207, 211–214
instruction, four-step method of, 221, 223
intellectual curiosity, 13
interactions, emotional
    comprehension of, 33
    Maslow's hierarchy of needs informing, 36
    role models of, 88–89
International Fire Service Training Association (IFSTA), xxxii
interpersonal communication, 76
intervention
    in command chain, 188
    corrective, 101
    counseling as, 175
    discretion with, 51
    by Powers, 249–250
    resistance to, 84
    with respect, 53
    size-up scheme for, 52
    timing of, 184, 247–248, 251
    unnecessary, 50
intrapersonal communication, 76
inventory, 241–242
investigation
    into bullying, 250
    after inspection, 219–221
investment, energy
    absentee leadership draining, 117–118
    of community, 96

into discipline, 95
intellectual curiosity as, 13
leader development regarding, xx, 70, 186–187
level of, 43
morale education regarding, 133
into problem-solving, 42–43
subordinates personal, 28

J

Jackson, Jesse, 15
Jersey City Fire Department, 132

K

K-12 saw, 217–218
Kant, Immanuel, 12
Kennedy, John F., xvii
Kibaki, Mwai, 37
King, Martin Luther, 45, 48
Kissinger, Henry A., xvii
kitchen, 233
    bullying in, 248
    rack, 87–89
    team building in, 173
Knot Gauntlet, 139
knowledge
    desire for, 13, 18
    levels of, xviii
Kroc, Ray, xvii

L

labor, division of, 147, 149
ladder pipe, aerial, 105, 107
Ladder 13, 237–243
Ladder 32, 237–243
Ladder 20, 217
Ladder 22, 128, 217–218
Ladder 222, xxxviii–xxxix
laissez-faire leadership, 29–30
laziness, leadership, 183
leader-coach, 172–173
leader-counselor
    as advisor, 174
    availability of, 198
    boundaries challenged for, 175
    counseling components used by, 176–178
    guidelines for, 178–179

skills honed by, 176
leadership. *See specific topics*
Lee, Robert E., 59
legitimate power, 63
Lemonie, Bill, 191
lesson plan, 232
life-support systems, 144–145
Lincoln, Abraham, 37, 58
loyalty, 1
    as action, 2
    to formal organization and subordinates, 54
    to socialized leader, 61

M

Macedonians, 254
maintenance
    of hardware, 170–171, 185
    issue discovered with, 240
    of personal protective equipment, xlvii
    of software, 171, 176, 185–186
    of work environment, 47, 49, 172
Malley, Kevin, xiv, xxxvii–xxxviii
management, xxiv, 47
    haphazard, 127
    motivation from, xl, 141
    study of, 17–18
    theory, xli
    transfer tool for, 187, 189
Manmouth County (NJ) Fire Academy, xxxviii
Martin, George R. R., 66
Maslow's hierarchy of needs, 35, 184, 199
    emotional interactions informed by, 36
    problems revealing, 40
maturity, emotional, 7, 60
Maxwell, John, 25
Mayo, George Elton, 136
media, 75
meetings, 213–214
    at-the-rig, 162
    on command philosophy change, 192
    communication opened by, 162, 206
    counseling session guidelines for, 178–179
    Finnegan holding, 191–192

observation at, 164
operations chief requirement for, xxvi
for orientation, 229, 233
preinspection, 207
with Strong, 203–206
as team-building session, 212
members, group
complaints from, 40
disenfranchisement of, 43
as experts, 29–30
fire service leader scrutinized by, 83
input from, 25–26, 30
trust of, 237
mental presence, xliii
mentorship
by Finnegan, xxxiii
opportunities for, 103, 184
primacy of, 228
problem-solving suggestion of, 197–198
Slip's story on, 196
as tool, 123
methods
four-step instruction, 221, 223
for mandates, xxvi
for supervision, 142
micromanagement, 47
minimalism, 101
Minneapolis Gas Company, 140–143
Miracle on the Hudson, 114
mission, 16–17
Montgomery, Bernard, xvii
mop, 248–249
morale
education on, 133–134
improvements in, 137
intangibility of, 135
socialized leadership improving, 61
speech on, 133
work environment impacted by, 135–136, 140–143
motivation
for achievement, xxxix
for confined space training, 144–145
Hawthorne experiments on, 136–140
from management, xl, 141
personalized and socialized power orientations as, 58
regarding subordinate development, 59
work environment impacted by, 140–143, 153
Murphy's law, 103

N

Nagurka, Frank, 109
needs, alignment of, xxiii
counseling for, 174, 200
discipline aiming for, 96, 188
with formal and informal organizations, 5
Maslow's hierarchy of needs informing, 36
between mission and personnel, 16
motivation from, 141, 143
by socialized leaders, 60
negative consequence, 103
negative discipline, 95–97, 188
counseling resulting in, 175
misdirected, 121
Neglia, Vincent, xxxiv
negligence, chain of, xlviii, 101
networking, 64
Newton, Johnny, 233–234
Night of Flames, xxvii
9/11 Memorial, 168
Nohria, Nitin, 87
nonverbal communication
for command, 26
for counseling, 177
verbal communication compared to, 77
North Hudson Regional Fire & Rescue (NHRF&R), xxvii
Barreres captain of, xxxi
dedication plaque at, 14
drill session safety checklist for, 11
Engine 13 and Battalion 3 of, xxxiv
high rise operations for, 27
personnel of, xxxii–xxxiii
Probationary Firefighter Journal for, 235
training schedule of, 10, 126

O

oath, 1, 33
obedience, 1, 8, 169

observation period, 120, 164
O'Driscoll, Kevin, 159
officer, company
   demonstration training regarding, 219–221
   as department bedrock, 71
   empowerment of, 225
   inspection involvement of, 207, 213
   as pivotal position, 70
   view of, 253
officer, engine, 158–160
officer, fire, 16
   accountability of, 122
   communication role modeling of, 88–89
   counseling challenge for, 170
   data dependence of, 80
   development of, 39, 149
   intervention regarding, 52–53
   qualities of, xlii
operating procedures, xxvi
operational plans, xlvii
operations, from theories, 18
operations chief, xxv–xxvi
opportunities
   from delegation, 152, 153
   discipline providing, 94
   for leader development, 205
   for mentorship, 103, 184
   problem-solving as, 40, 41
   from promotion, 165
orders
   by casual leaders, 111
   exponential factor of, 112
   for gloves, 180–181
   justifiable, 158, 187
   obedience to, 1, 169
   respect for, 13
   value of, 143
organization, xlviii, 8, 12, 31
   planning for, 9
   safety checklist for, 11
   training schedule for, 10
organizations, formal
   bullying consequences in, 247, 250–251
   decay of, 106, 112
   emotional maturity of, 7
   engagement rules from, 4
   expectations of, 93, 94–95
   fire department as, 3–5
   informal leader influence on, 35
   informal organization integration with, 4–5, 34, 40
   leadership failure of, 118
   loyalty to, 54
   power misuse disrupting, 57–58
   respect for, 14
   tools of, 155
organizations, informal, 3
   command vacuum filled by, 21
   formal organization integration with, 4–5, 34, 40
   responsibility entanglement with, 54–55
orientation
   for Asher, 227–232, 234
   checklist for, 231–232
   comparison of, 234
   greeting for, 227, 230
   meetings for, 229, 233
   for Newton, 233–234
   primacy regarding, 228
   veteran buddy system for, 229, 231
Orio, Mike De, 151
ownership, 3
   of commander, 204, 209
   productivity enhanced by, 138
   subordinates allowed, 19–20, 143

P

paralanguage, 86
parent ego state, 86
participative leadership, 28–29
Patton, George, 37–38
Pelosi, Nancy, 12
people skills, xxv
performance, xiii, 69
personal protective equipment (PPE), xlvii
personalized power orientation
   coercion of, 59
   self-glorification of, 58
   socialized power orientation compared to, 58–62
Peter principle, 68–71
Plan of the Day, 128
planning, xlviii, 31, 232
   ambush leader missing, 125
   casual and sensual leadership missing, 110
   chief lack of, 128–129
   coaching requiring, 173

coordination included in, 12
imagination failure for, 104
for informal training sessions, 220
for organization, 9
preinspection meeting for, 207
situation-specific action, 17
Plato, 15, 58, 175
population, xxvii–xxviii
position, leadership
fulfillment of, xx
pivotal, 70
privilege from, 38
servant-leader maintaining, 20
positive consequence, 103
positive discipline, 93–95, 188
Potter, Alexandra, 101
power. *See also* personalized power orientation; socialized power orientation
of fire service leader communication, 81–84
formal, 62–63
informal, 63, 65
as nonrenewable resource, 57
proactive and reactive, 71–72
types of, 63–65, 66–68, 199
work environment poisoned by, 57–58
Powers, Grace
bullying intervention of, 249–250
communication by, 212, 214, 219
confidence challenge of, 203, 209, 211, 214
inspections by, 217–218, 219–221
leadership guidance for, 204
remedial actions by, 239–241
shift change to, 238–239
skills list of, 213–214, 225
Strong meeting with, 203–206
practice
for communication, 82
of organization, 9
in soft environment, 113
preparation, xxi
for deadline, 128–129
for decision-making, 81
meeting regularity for, 162
preinspection, 207, 211–214
for promotional exam, 69
for soft environment communication, 85
pride, 145
primacy, 161, 228

Preinspectionectives, 3, 147, 188
prioritization, 9, 40
privilege, 38–39
proactive leadership, 50, 160
probationary firefighter, xxx–xxxi
bullying of, 246–247
importance of, 248
journal for, 235
problem-solving, 39
absentee leadership avoiding, 118
communication regarding, 42, 84, 88–89
for counseling, 177, 178–179
energy investment into, 42–43
feelings challenge for, 133
functional fixity inhibiting, 104
leadership skills for, 225
mentorship suggested for, 197–198
as opportunity, 40, 41
personalities separated in, 174
proactivity for, 41–42, 50, 55
productivity
absentee leadership opposite of, 117
ambush leadership opposite of, 124
ownership enhancing, 138
work environment changes regarding, 136–137
profession, of firefighting, 167, 169
promotion, 2. *See also* exam, promotional
to 1st Platoon, 163–164
opportunities from, 165
peter principle for, 68–71
protocols, 96–97
public communication, 76
punishment
hierarchy of, 95
transfer as, 188–189

Q

quality, leadership
commitment to, xxiii, xli
leadership quantity compared to, xxii
quantity, leadership, xxii
Quidor, Steve, 37, 54, 73

R

radio, 120–121
rank
    influence of, xlii
    merit for, 48
    privilege from, 38
    problem-solving regarding, 42
    promotional exam advancing, 69
reactive leadership, 160
ready status
    as department expectation, 161
    meetings confirming, 162
    remedial actions regarding, 239–241
    transfer for, 218
recognition, xxix–xxx
    employees influenced by, 138, 139
    of Slip, 193–194, 198
    of subordinates, 42–43
referent power, 65, 67
regionalization, xxvi
reinforcement
    affirmative, 94
    by chief, 122
    of formal organization expectations, 94–95
relationships, work, 17
    camaraderie in, 45–46
    cultivation of, 8
    between leader and subordinates, xvii, xxxix, 55, 66, 88
    mentorship for, 197–198
    motivation considering, 138
    skill portfolio for, 182, 199
    socialized leadership dependent on, 60
reports, xlviii, 31, 96
    of bullying, 249–250, 252
    false, 241–242
    on training, 219
reputation, 53, 97, 99, 243
requirements
    comprehension of, 178
    for meetings, xxvi
    for radio, 120–121
    for subordinates, 73
resources
    on leadership, xli, xlvi
    nonemergency drain of, 83
    power as, 57
respect
    for formal organization, 14
    from informal power, 65
    intervention with, 53
    leader earning, 38
    for orders, 13
responsibility, xiii, xl, 205
    bottom-up and top-down, 1
    for casual attitudes, 159
    of chief, 70, 118
    over comfort zone, 53
    of command, 204
    of department, 122
    as emotional maturity, 7
    engine officer identifying, 160
    of fire service leaders, xliii
    of formal organization, 94–95
    informal organization entanglement with, 54–55
    leader-counselor understanding of, 178
    privilege outweighed by, 39
    problem-solving as, 42
    of subordinate development, 148
Revis, Beth, 59
reward power, 63
role model, 19
    chief as, 118–119
    of communication, 88–89
roles, leadership, 34–35
Roof Pack, 142

S

safety
    certification of, 240
    department responsibility for, 122
    organization checklist for, 11
scenarios
    autocratic leadership required by, 26, 67
    communication portfolio for, 81
    hypothetical, xliv
    laissez-faire leadership required by, 30
    of leadership failure, 105–107
    real-life, xiii, xxiv
    transactional analysis clarifying, 86–87
schedule, 9
    ambush leader disregard for, 125
    fraternization-free time in, 55
    for North Hudson Regional Fire & Rescue training, 10, 126

Index

Schwarzkopf, Herbert Norman, 117
Schweitzer, Albert, 59
security, 36, 140
self-assessment
　facilitation of, 199
　of leader-counselor, 178
　by Powers, 203
self-discipline, 93
self-esteem, 36
self-glorification, 58
self-realization, 36, 94
seniority, 191–192, 194
sensual leadership, 109
　casual leadership in tandem with, 110
　confusion from, 114
　identification of, 111
separation, of leader, 55
"The Servant as Leader" (Greenleaf), 20
servant-leader, 20–21
　chief as, 113
　socialized leadership compared to, 60, 61
service position, xxxi, 18
Shaw, George Bernard, 73
shift changeover, 238–239
size-up scheme, 52
skill portfolio, xlii, 81
　for communication, 83, 87
　orientation using, 231
　in preinspection meeting, 213–214
　for work relationships, 182, 199
Slip, Andrew, 191
　apology by, 195, 196
　change difficulty of, 194
　counseling session with, 193–198
　mentorship story by, 196
　outburst of, 192
small-group communication, 76
social needs, 36
socialized leader, 60–61
socialized power orientation
　authority use with, 60
　personalized power orientation compared to, 58–62
　work environment improved by, 61
soft environment, xliv
　communication in, 81–83, 84–85
　hard environment impact from, 68, 113, 205
　hard-think for, xlvii–xlviii
　interpersonal conflict resolution in, 50
　leadership practice in, 113
　mind-sets, xlv
　360-degree firefighter in, 245
soft-think, xlv, xlviii
software
　as asset, 230
　coaching and counseling for, 172
　maintenance of, 171, 176, 185–186
specialization, 149–150
spot-counseling, 176
Squad 7, 233–234
staff, xlviii, 31
standardized operating procedures (SOPs), 3
　for apparatus changeover, 241
　autocratic leadership codified by, 68
　emergencies addressed with, 72
　leadership failure without, 105
　radio requirement of, 120–121
　structure established by, 27
states, leadership, 48
status quo
　of formal leader, 34
　interpersonal conflict as, 46, 49
stewardship, 20
strategy, 74, 81
Strong
　leadership guidance by, 208–209
　meeting with, 203–206
study groups, xliv, 133
styles, leadership, xlii
　autocratic, 25–28, 67, 68
　combination of, 25, 31
　democratic, 30–31
　laissez-faire, 29–30
　participative, 28–29
　power types and, 66–68
　variety of, 15
subordinates, 39. *See also* development, subordinate
　ambush leadership disgruntling, 125
　comfort zone regarding, 38
　communication requirements for, 73
　counseling session worksheet for, 201
　empowerment of, 19, 149–150
　energy investment of, 28

expectations for, 155
expert power assigned by, 64
fire service leader defined by, 117
leader relationship with, xvii, xxxix, 55, 66, 88
leadership failure regarding, 148
loyalty to, 54
motivation induced for, 141, 143
ownership allowed to, 19–20, 143
problems presented by, 40
readiness of, 171
recognition of, 42–43
referent power assigned by, 65
resistance of, 59
Sun Tzu, 20
Sundaram, Meenakshi, 101
supervision, 42, 134, 147–148, 151, 165
   freelancing minimized with, 115
   hyperawareness for, 51
   leadership failures of, 47, 107, 112, 121, 249
   methods for, 142
support, mutual, 109
   of fire officers, 53
   of formal organization, 4
synergy, 167–168, 170
synthesis, xviii

T

Tactical Perspectives DVD series team, 110
team-building, 212, 214
technology, 79
termination, 96
thought, leadership, 101
threat, 249
   charges from, 250
   to command, 51
3C model, 82–83
360-degree check, 160
360-degree commitment, 252
360-degree firefighter, 245–246
timing
   inappropriate behavior regarding, 251
   of interpersonal conflict, 50, 52
   of intervention, 184, 247–248, 251
   orchestration of, 193
title, formal, 16

tool, leadership
   communication skills as, 82
   creativity as, 141–142
   discipline as, 91–92, 94, 99
   of formal organizations, 155
   mentorship as, 123
   in preinspection meeting, 213–214
   reactive power as, 71–72
   transfer as, 187, 189
   for transformation, 19–20
tragedies, 41
   autocratic leadership preventing, 67
   from casual leadership, 114
   communication failures resulting in, 83
   discipline preventing, 53
   leadership failure resulting in, 105–107
   thoughtless leadership creating, 102
training, xxxii
   communication exercise for, 89
   confined-space, 144–145
   demonstrations regarding, 219–221
   hard-think in, xlvi
   North Hudson Regional Fire & Rescue schedule for, 10, 126
   orientation as, 227–232
   as positive discipline, 94
   for radio procedures, 120
   reports on, 219
   as socialized leadership, 61
   on work relationships, 182
transactional analysis, 87–89, 184
   components of, 85
   ego states of, 86
transfer
   decision-making for, 181, 183–184
   discipline compared to, 188
   laziness of, 183
   as leadership failure, 182–183, 189
   as punishment, 188–189
   for ready status, 218
   as tool, 187, 189
trust, 72
   from command presence, 164
   consistency gaining, 6
   delegation exercise for, 151–152
   of group members, 237
   orientation building, 231

violation of, 241
Turner, Mayor, 62
Twain, Mark, 85, 131

U

unification
    of command, 52
    democratic leader creating, 30
    of department, 156
    of 1st Platoon, 165
    operations chief cajoling, xxv–xxvi
    transfer punishment against, 189
unintended consequences, 103, 243
United States fire service, xix, xlii

V

Van Valkenburg, George, 37
Vasta, Frank, 54
verbal command, 26
verbal communication, 76–77
veteran firefighter, 229, 231
victimization, 106
violations, of rules, 7–8
vision
    communication of, 6, 159–160
    for 1st Platoon, 164
    levels of, 253–254
    into reality, xl
visual communication, 78–79

W

watches, 134–135
We All Shine On, xxxiv
weak-kneed leadership
    crew comfort prioritized by, 21–22
    ramifications of, 68
    socialized leadership compared to, 60
    unacceptability of, 23
Weehawken Fire Department (WFD), xxx, 62
    command of, 16
    personnel on, xxxii
    Roof Pack of, 142
Weisman, Avery, 131
Wilde, Oscar, 131
Wilkinson, Bud, 131
windows, 144

work. *See* assignments, work; environment, work; relationships, work
World Wide Web, 78
written communication, 78